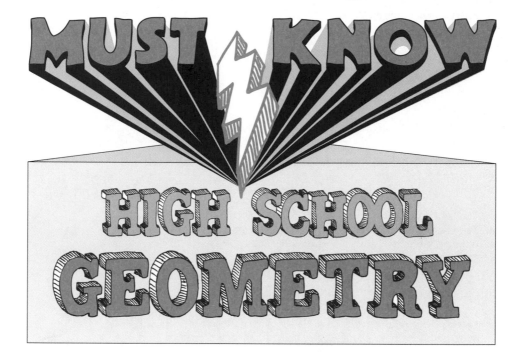

Second Edition

# Allen Ma

# Amber Kuang

New York   Chicago   San Francisco   Athens   London   Madrid
Mexico City   Milan   New Delhi   Singapore   Sydney   Toronto

1 2 3 4 5 6 7 8 9   LCR   27 26 25 24 23 22

ISBN        978-1-264-28614-0
MHID        1-264-28614-7

e-ISBN      978-1-264-28615-7
e-MHID      1-264-28615-5


Interior design by Steve Straus of Think Book Works.
Cover and letter art by Kate Rutter.


McGraw Hill books are available at special quantity discounts to use as premiums and sales promotions or for use in corporate training programs. To contact a representative, please visit the Contact Us pages at www.mhprofessional.com.


McGraw Hill is committed to making our products accessible to all learners. To learn more about the available support and accommodations we offer, please contact us at accessibility@mheducation.com. We also participate in the Access Text Network (www.accesstext.org), and ATN members may submit requests through ATN.

## Dedication

We would like to dedicate this book to Lorraine Poppe, Madeline Donahoe, and Bill Ma. We are forever grateful for the endless encouragement and inspiration you have provided us. Thank you for believing in us before we believed in ourselves. You truly taught us the **must knows** of being a great educator.

## Authors' Acknowledgments

This book would never have been possible without the support of so many people. We would like to thank our family members: Steven, Vivian, Sherry, Andrew, Jonathan, Linda, Fred, Carol, Gordon, Gavin, Brandon, Zachary, Fara, Sandi, and Mike, for their constant love and encouragement. We would like to extend our gratitude to Michael Isoldi and our agent Grace Freedson for giving us this amazing opportunity. Thank you also to Daina Penikas, Tama Harris, and Rishabh Gupta for their contributions to this book. Finally, we would like to thank our Kennedy High School family. We are so proud to be part of such an incredible learning community with supportive administrators, teachers, families, and students.

# Contents

# Introduction

**W**elcome to your new geometry book! Let us try to explain why we believe you've made the right choice with this new edition. This probably isn't your first go-round with either a textbook or other kind of guide to a school subject. You've probably had your fill of books asking you to memorize lots of terms. This book isn't going to do that—although you're welcome to memorize anything you take an interest in. You may also have found that a lot of books make a lot of promises about all the things you'll be able to accomplish by the time you reach the end of a given chapter. In the process, those books can make you feel as though you missed out on the building blocks that you actually need to master those goals.

With *Must Know High School Geometry,* we've taken a different approach. When you start a new chapter, right off the bat you will immediately see one or more **must know** ideas. These are the essential concepts behind what you are going to study, and they will form the foundation of what you will learn throughout the chapter. With these **must know** ideas, you will have what you need to hold it together as you study, and they will be your guide as you make your way through each chapter, learning how to build new knowledge and skills on them.

To build on this foundation, you will find easy-to-follow discussions of the topic at hand, and these are accompanied by comprehensive examples that show you how to apply what you're learning to solve typical geometry questions. Each chapter ends with review questions—more than 300 throughout the book— that are designed to instill confidence as you practice your new skills.

This book has other features that will help you on this geometry journey of yours. It has a number of sidebars that will provide helpful information or just serve as a quick break from your studies. The **BTW** sidebars ("by the way") point out important information as well as study tips and exceptions to the rule. Every once in a while, an **⊕IRL** sidebar ("in real life") will tell you what you're studying has to do with the real world; other IRLs may just be interesting factoids.

But that's not all—this new edition has taken it a step further. We know our geometry students well and we want to make sure you're getting the most out of this book. We added new **EASY MISTAKE** sidebars that point out common mistakes and things *not* to do. For those needing a little assistance, we have our **EXTRA HELP A+** feature, where more challenging concepts, topics, or questions are given some more explanation. And finally, one special note for the teachers (because we didn't forget about you!)—a **Teacher's Guide** section at the back of the book is a place where you can go to find tips and strategies on teaching the material in the book, a behind-the-scenes look at what the authors were thinking when creating the material, and resources curated specifically to make your life easier!

In addition, this book is accompanied by a flashcard app that will give you the ability to test yourself at any time. The app includes more than 100 "flashcards" with a review question on one "side" and the answer on the other. You can either work through the flashcards by themselves or use them alongside the book. To find out where to get the app and how to use it, go to "The Flashcard App."

Now that you're ready to get started, let us introduce you to your guides throughout this book. Between them, Allen Ma and Amber Kuang teach algebra, geometry, trigonometry, precalculus, calculus, and discrete math. They have a clear idea about what you should get out of a geometry course and have developed strategies to help you get there. They also have seen the kinds of trouble that students can run into, and they are experienced hands at solving those difficulties. In this book, they apply that experience both to showing you the most effective way to learn a given concept as well as how to extricate yourself from traps you may have fallen into. They will be trustworthy guides as you expand your geometry knowledge and develop new skills.

Before we leave you to your authors' capable guidance, let us give you one piece of advice. Although we know that saying something "is the *worst*" is a cliché, if anything in geometry *is* the worst, it's formal geometric proofs. Let your new teachers introduce you to them and show you how to apply them confidently to your geometry work. Mastering geometric proofs will give you an invaluable advantage for the rest of your math career.

Good luck with your studies!

*The Editors at McGraw Hill*

# The Flashcard App

This book features a bonus flashcard app. It will help you test yourself on what you've learned as you make your way through the book (or in and out). It includes 100-plus "flashcards," with information and terms on both the "fronts" and "backs" of the cards. You can use the app in two ways: (1) You can jump right into the app and start from any point that you want. Or (2) you can take advantage of the handy QR codes near the end of each chapter in the book; they will take you directly to the flashcards related to what you're studying at the moment.

To take advantage of this bonus feature, follow these easy steps:

Search for **McGraw Hill Must Know** App from
either Google Play or the App Store.

↓

Download the app to your smartphone or tablet.

↓

Once you've got the app,
you can use it in either of two ways.

↙          ↘

Just open the app and
you're ready to go.

Use your phone's QR Code reader
to scan any of the book's QR codes.

You can start at the beginning,
or select any of the chapters listed.

You'll be taken directly to the flashcards
that match your chapter of choice.

↘          ↙

**Be ready to test your
geometry knowledge!**

# MUST KNOW

- Terms and definitions in geometry provide the foundation for proving geometric theorems.

- Geometric theorems enable us to develop the statements and reasons necessary to write a formal geometric proof.

- The reflexive property tells us that an angle or segment is congruent to itself.

astering geometry requires the understanding of many properties and definitions. This chapter will focus on the different vocabulary that weaves its way through many geometry units. You might find it helpful to create index cards to help commit the information to memory. It is also critical to become familiar with the different symbols that are used to represent some of the vocabulary words.

## The Basics

Let's begin by defining an **angle**. Represented by $\angle$, an angle is the space between two intersecting lines, line segments, or rays. Angles can be named using one letter if it is the only angle at that location, or three letters, the middle letter being the vertex of the angle. The accompanying diagram shows an angle that would be written as $\angle ABC$, where $B$ is the vertex of the angle.

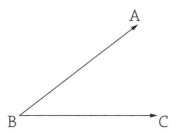

Another symbol we frequently see in geometry is the symbol for a **line**, $\leftrightarrow$. The symbol for a line has arrows on both sides to show that it extends infinitely in both directions. The accompanying diagram shows a line that would be written as $\overleftrightarrow{AB}$.

A **ray** begins at an endpoint and then continues infinitely in one direction. Therefore, a ray is represented by the symbol →. The accompanying diagram shows a ray that would be written as $\overrightarrow{CD}$.

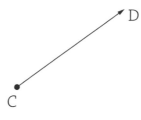

A **segment** has definite endpoints on both sides, which is why it is represented by the symbol —. A segment does not have arrows indicating that it is a part of a line with definite endpoints on both sides. The accompanying diagram shows a segment that would be written as $\overline{EF}$.

## Bisectors and Midpoints

Now that we are more familiar with some of the symbols that will appear throughout this book, we can begin learning some of the important geometric terms. Let's begin by learning about **bisectors**. There are two different types of bisectors: angle bisectors and segment bisectors. A bisector divides an angle or segment into two congruent parts. This means that an **angle bisector** divides an angle into two congruent angles, and a **segment bisector** divides a segment into two congruent segments. In geometry, the symbol for congruence is ≅. When a segment is bisected, the point of intersection is called the **midpoint** of the segment. This means that the midpoint of a segment also divides a segment into two congruent segments.

Let's look at some different problems that involve bisectors. Our first bisector example will help us understand how to develop information for a geometric proof.

> **BTW**
>
> Read problems carefully. If a question says $\overline{AB}$ is a bisector, that does not mean that $\overline{AB}$ is being divided in half. It means $\overline{AB}$ is dividing something else in half.

In the accompanying diagram, $\overline{CD}$ bisects $\overline{AB}$ at $E$. What can we conclude from this given information?

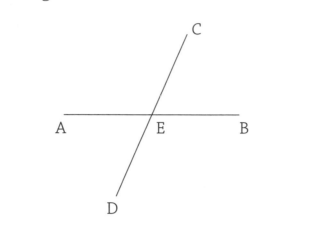

This question states that $\overline{AB}$ is the segment being bisected. This means $\overline{AB}$ will be divided into two congruent segments that make $E$ the midpoint of $\overline{AB}$. Therefore, $\overline{AE} \cong \overline{EB}$. See the markings in the accompanying diagram that show $\overline{AE} \cong \overline{EB}$.

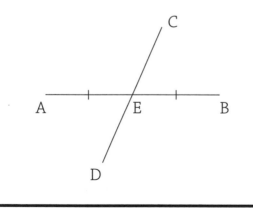

**EASY MISTAKE**

Do NOT conclude that $\overline{CE}$ is congruent to $\overline{DE}$. $E$ is NOT necessarily the midpoint of $\overline{CD}$.

Did you know it is possible for the bisector to be bisected? Let's look at an example.

▶ In the accompanying diagram, $\overline{AB}$ and $\overline{CD}$ bisect each other at $E$. What can we conclude from this given information?

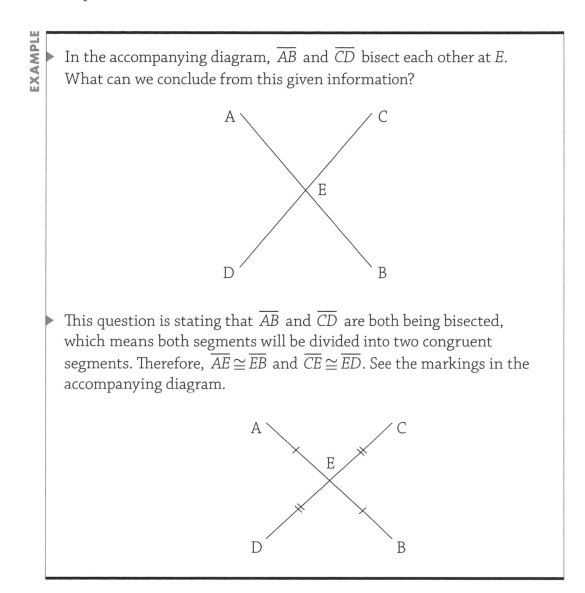

▶ This question is stating that $\overline{AB}$ and $\overline{CD}$ are both being bisected, which means both segments will be divided into two congruent segments. Therefore, $\overline{AE} \cong \overline{EB}$ and $\overline{CE} \cong \overline{ED}$. See the markings in the accompanying diagram.

It is also important to see how bisectors can show up in problems that require algebraic solutions.

▶ In the accompanying diagram, $\overline{DE}$ bisects $\overline{AM}$ at $T$. $AT = x + 10$ and $TM = 5x - 2$. Find the length of $AT$.

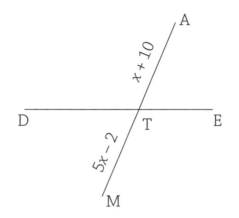

▶ This question is stating that $\overline{AM}$ is being bisected, which means $\overline{AM}$ will be divided into two congruent segments. Therefore, $\overline{AT} \cong \overline{TM}$. To solve this algebraically, we will first set the two segments equal and solve for $x$. We can then substitute the value of $x$ into the expression representing the length of $AT$.

$$AT = TM$$
$$x + 10 = 5x - 2$$
$$\underline{-x + 2 \qquad -x + 2}$$
$$\frac{12}{4} = \frac{\cancel{4}x}{\cancel{4}}$$
$$3 = x$$
$$AT = x + 10$$
$$AT = 3 + 10$$
$$AT = 13$$

Now that we have seen how the term *bisector* is applied to segments, let's apply this same concept to angles. Our first example will help you develop information for a geometric proof.

▶ In the accompanying diagram, $\overline{AD}$ bisects $\angle CDE$. What can we conclude from this given information?

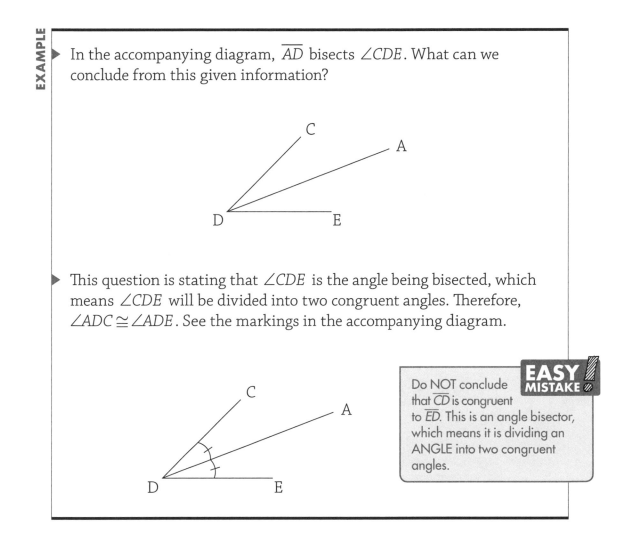

▶ This question is stating that $\angle CDE$ is the angle being bisected, which means $\angle CDE$ will be divided into two congruent angles. Therefore, $\angle ADC \cong \angle ADE$. See the markings in the accompanying diagram.

**EASY MISTAKE**

Do NOT conclude that $\overline{CD}$ is congruent to $\overline{ED}$. This is an angle bisector, which means it is dividing an ANGLE into two congruent angles.

We can also take an algebraic approach toward angle bisectors, as you can see in the accompanying example.

In the accompanying diagram, $\overline{AE}$ bisects $\angle DEF$, $m\angle DEF = 10x + 4$, and $m\angle AED = 3x + 12$. Find $m\angle DEF$.

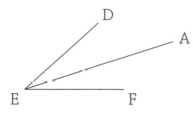

$\angle DEF$ is being bisected, which means it is being divided into two congruent angles. Therefore, $\angle AED \cong \angle AEF$. See the markings in the accompanying diagram. If the angles are congruent, and $m\angle AED = 3x + 12$, that also means that $m\angle AEF = 3x + 12$. We can add these two angles together and make their total equal to $m\angle DEF$.

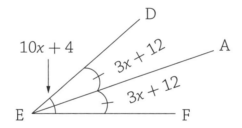

$$m\angle AED + m\angle AEF = m\angle DEF$$
$$3x + 12 + 3x + 12 = 10x + 4$$
$$\cancel{6}x + 24 = 10x + \cancel{4}$$
$$\underline{-\cancel{6}x - 4 \quad -6x - \cancel{4}}$$
$$\frac{20}{4} = \frac{\cancel{4}x}{\cancel{4}}$$
$$5 = x$$
$$m\angle DEF = 10x + 4$$
$$m\angle DEF = 10(5) + 4$$
$$m\angle DEF = 54°$$

▶ A second method can be used to answer this question. Because a bisector divides an angle into two congruent angles, $2(m\angle AED) = m\angle DEF$.

$$2(3x + 12) = 10x + 4$$
$$6x + 24 = 10x + 4$$
$$\underline{-6x \quad -4 \quad -6x - 4}$$
$$\frac{20}{4} = \frac{4x}{4}$$
$$5 = x$$
$$m\angle DEF = 10x + 4$$
$$m\angle DEF = 10(5) + 4$$
$$m\angle DEF = 54°$$

## Types of Angles

This section of the chapter will focus on important angle definitions. This is where we will learn about right angles, vertical angles, complementary angles, supplementary angles, and angles that form a linear pair. These relationships will help us solve algebraic problems and geometric proofs.

When two lines intersect, they form congruent **vertical angles**. You can see in the accompanying diagram that $\overline{AB}$ and $\overline{CD}$ intersect at $E$. These intersecting segments form two pairs of congruent vertical angles, $\angle AEC \cong \angle BED$ and $\angle AED \cong \angle BEC$.

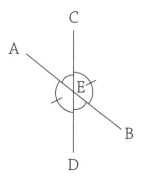

Let's look at how this can appear in a problem that requires an algebraic solution.

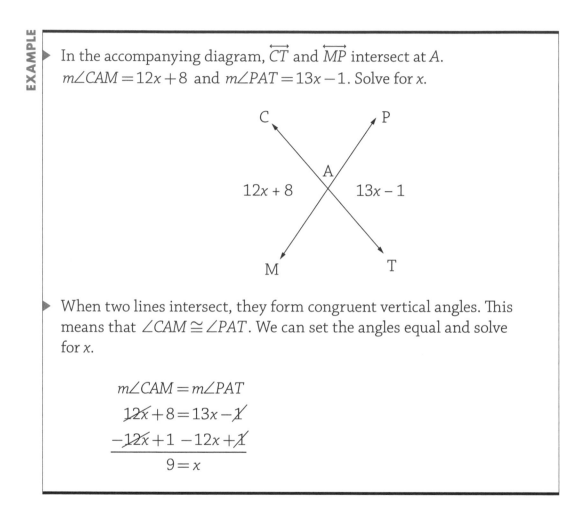

EXAMPLE

▶ In the accompanying diagram, $\overleftrightarrow{CT}$ and $\overrightarrow{MP}$ intersect at $A$. $m\angle CAM = 12x + 8$ and $m\angle PAT = 13x - 1$. Solve for $x$.

▶ When two lines intersect, they form congruent vertical angles. This means that $\angle CAM \cong \angle PAT$. We can set the angles equal and solve for $x$.

$$m\angle CAM = m\angle PAT$$
$$12x + 8 = 13x - 1$$
$$\underline{-12x + 1 \quad -12x + 1}$$
$$9 = x$$

**Perpendicular lines**, represented by the symbol $\perp$, form **right angles**. All right angles equal 90°, which means all right angles are congruent to one another. Look at the accompanying diagram that shows $\overline{AB}$ perpendicular to $\overline{CD}$. The right angles are represented by placing a box at E. Here we can conclude that $\angle AEC$, $\angle AED$, $\angle BEC$, and $\angle BED$ are right angles. We can also conclude that $\angle AEC \cong \angle AED \cong \angle BEC \cong \angle BED$, as seen in the accompanying diagram.

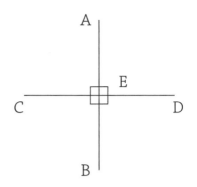

The accompanying example shows how this same concept can be applied to a problem that requires an algebraic solution.

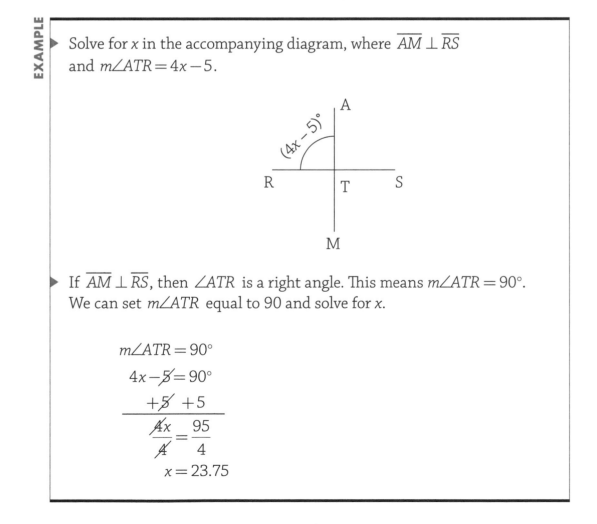

**EXAMPLE**

▶ Solve for $x$ in the accompanying diagram, where $\overline{AM} \perp \overline{RS}$ and $m\angle ATR = 4x - 5$.

▶ If $\overline{AM} \perp \overline{RS}$, then $\angle ATR$ is a right angle. This means $m\angle ATR = 90°$. We can set $m\angle ATR$ equal to 90 and solve for $x$.

$$m\angle ATR = 90°$$
$$4x - \cancel{5} = 90°$$
$$\underline{+\cancel{5}\ +5}$$
$$\frac{\cancel{4}x}{\cancel{4}} = \frac{95}{4}$$
$$x = 23.75$$

Many geometric problems incorporate more than one concept in a single question. The accompanying problem is a good example of this.

▶ Use the accompanying diagram and the given information to answer this question. $\overline{AB}$ is the perpendicular bisector of $\overline{CD}$ at E. What can you conclude from this given information?

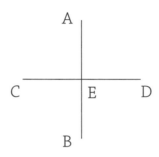

▶ This question is stating that $\overline{CD}$ is being bisected. This means $\overline{CD}$ will be divided into two congruent segments. Therefore, $\overline{CE} \cong \overline{ED}$, as shown in the diagram below. The question also states that $\overline{AB}$ is perpendicular to $\overline{CD}$, which means they form right angles at point E. Therefore, $\angle AEC$, $\angle AED$, $\angle BEC$, and $\angle BED$ are right angles. It can also be concluded that $\angle AEC \cong \angle AED \cong \angle BEC \cong \angle BED$. The accompanying diagram shows these conclusions.

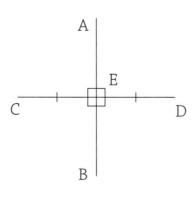

**Complementary angles** are a pair of angles whose sum is 90°. If two angles add to 90°, we call one angle the *complement* of the other angle. **Supplementary angles** are a pair of angles that sum to 180°. If two angles add to 180°, we call one angle the *supplement* of the other angle. When two adjacent angles form a **straight angle**, a line, they are called **linear pairs.** A straight angle is equal to 180°, which means that two angles that form a linear pair are supplementary angles.

The accompanying example shows how the definition of complementary angles is used to algebraically solve for the measure of an angle.

▶ Given two complementary angles, one of the angles is 9 less than twice the other angle. What is the measure of the larger angle?

▶ Complementary angles are two angles whose sum is 90°. If we can create expressions representing the measure of each angle, we can add them together and set them equal to 90° to solve for $x$. We can then substitute the value of $x$ into each expression to find the measure of each angle.

▶ Let $x =$ The first angle.

▶ Let $2x - 9 =$ The second angle.

$$x + 2x - 9 = 90$$
$$3x - 9 = 90$$
$$\underline{+9 \quad +9}$$
$$\frac{3x}{3} = \frac{99}{3}$$
$$x = 33$$

▶ The first angle, $x$, is 33°. The second angle, $2x - 9$, is found by plugging in 33 for $x$. $2(33) - 9 = 57°$. The larger angle is 57°.

In our classes, whenever there are questions that have diagrams with angles, we always tell our students to look for angles that form a line. We know angles that form a line add to 180°, which makes it easier to find the measures of the other angles. The following example shows this concept.

▸ In the accompanying diagram of $\overleftrightarrow{AEB}$, $m\angle AEC = 4x - 2$ and $m\angle CEB = x + 12$. Solve for $x$.

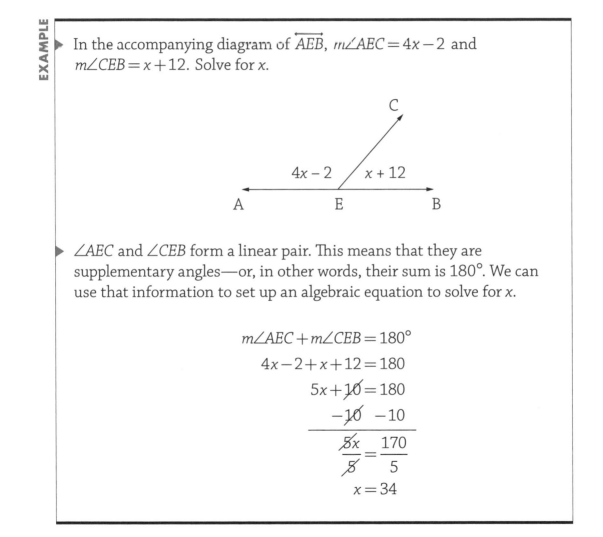

▸ $\angle AEC$ and $\angle CEB$ form a linear pair. This means that they are supplementary angles—or, in other words, their sum is 180°. We can use that information to set up an algebraic equation to solve for $x$.

$$m\angle AEC + m\angle CEB = 180°$$
$$4x - 2 + x + 12 = 180$$
$$5x + 10 = 180$$
$$-10 \quad -10$$
$$\frac{5x}{5} = \frac{170}{5}$$
$$x = 34$$

# Reflexive, Substitution, and Transitive Properties

A large portion of geometry focuses on formal geometric proofs. An entire chapter of this book is devoted to formal proofs, but before we can dive into that chapter, there are important properties and postulates we need to get under our belt. Let's begin by learning about the **reflexive property**. The reflexive property states that an angle or segment is always congruent to itself. This property is helpful to us because there are times where one angle or segment is a part of two different scenarios within a diagram. We will use the accompanying diagram of $\overline{ABCD}$ to show this property.

As you can see in the diagram, points $B$ and $C$ form $\overline{BC}$ on $\overline{AD}$ in such a way that $\overline{BC}$ is a part of $\overline{AC}$ and $\overline{BC}$ is also a part of $\overline{BD}$. The accompanying diagram will make that concept clearer.

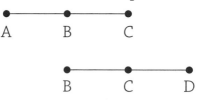

The reflexive property says that the $\overline{BC}$ that is a part of $\overline{AC}$ is congruent to the $\overline{BC}$ that is a part of $\overline{BD}$ because it is the same $\overline{BC}$. Therefore, we can conclude that $\overline{BC} \cong \overline{BC}$.

We can use the reflexive property with angles as well. In the accompanying diagram of $\triangle ABC$, $\angle B$ is an angle in $\triangle ABC$ and an angle in $\triangle DBE$. The reflexive property says that the $\angle B$ in $\triangle ABC$ is congruent to the $\angle B$ in $\triangle DBE$. Therefore, we can conclude that $\angle B \cong \angle B$.

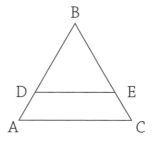

Two other properties we must also mention are the substitution property and the transitive property. The **substitution property** says that if $a = b$ in a question, then any $b$ in the question can be rewritten as $a$. The **transitive property** states that if $a = b$ and $b = c$, then $a = c$. This property might make more sense if we use it with a more logical example. If Lorraine is in Jerry's English class and Jerry is in Madeline's English class, then we can conclude, by the transitive property, that Lorraine must be in Madeline's English class.

## Addition and Subtraction Postulates

There are times when we may be missing a piece of an angle or segment that is needed to complete a proof, or we would like to get rid of a piece of an angle or segment to complete a proof. We can add and subtract segments and angles from one another in geometric proofs using the **addition** and **subtraction postulates**. The addition postulate states that if we add two congruent segments or angles to two segments or angles that are already congruent, the results will also be congruent. The subtraction postulate states that if we subtract two congruent segments or angles from two segments or angles that are already congruent, the results will also be congruent.

We will use the following diagram of $\overline{MATH}$ with $\overline{MA} \cong \overline{TH}$ to show how we could use the addition postulate to prove $\overline{MT} \cong \overline{HA}$.

The reflexive property says that $\overline{AT} \cong \overline{AT}$, meaning that the $\overline{AT}$ that is a part of $\overline{MT}$ is congruent to the $\overline{AT}$ that is a part of $\overline{HA}$.

We can see in the diagram that $\overline{MA} + \overline{AT} = \overline{MT}$ and that $\overline{HT} + \overline{AT} = \overline{HA}$. Therefore, $\overline{MT} \cong \overline{HA}$ from the addition postulate.

It is important that we see this concept with angles as well. The following example will apply the subtraction postulate to angles. We will be setting up this example as a formal proof. Normally in a formal proof we organize our information in a table with the statements we can make on one side and the reasons for those statements on the other side.

EXAMPLE

In the accompanying diagram of $\overleftrightarrow{ABC}$, $\angle ABE \cong \angle CBD$. Prove $\angle ABD \cong \angle CBE$.

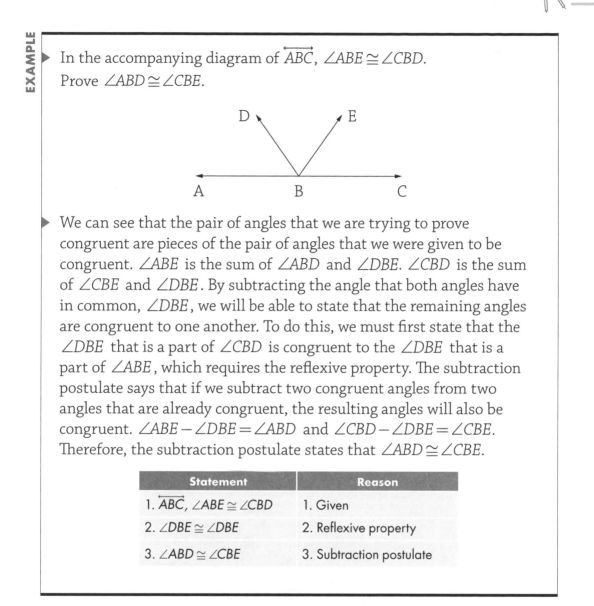

We can see that the pair of angles that we are trying to prove congruent are pieces of the pair of angles that we were given to be congruent. $\angle ABE$ is the sum of $\angle ABD$ and $\angle DBE$. $\angle CBD$ is the sum of $\angle CBE$ and $\angle DBE$. By subtracting the angle that both angles have in common, $\angle DBE$, we will be able to state that the remaining angles are congruent to one another. To do this, we must first state that the $\angle DBE$ that is a part of $\angle CBD$ is congruent to the $\angle DBE$ that is a part of $\angle ABE$, which requires the reflexive property. The subtraction postulate says that if we subtract two congruent angles from two angles that are already congruent, the resulting angles will also be congruent. $\angle ABE - \angle DBE = \angle ABD$ and $\angle CBD - \angle DBE = \angle CBE$. Therefore, the subtraction postulate states that $\angle ABD \cong \angle CBE$.

| Statement | Reason |
|---|---|
| 1. $\overleftrightarrow{ABC}$, $\angle ABE \cong \angle CBD$ | 1. Given |
| 2. $\angle DBE \cong \angle DBE$ | 2. Reflexive property |
| 3. $\angle ABD \cong \angle CBE$ | 3. Subtraction postulate |

## EXTRA HELP

In this formal proof, students may have difficulty deciding if the addition or subtraction postulate should be used. $\angle ABE$ is larger than $\angle ABD$ and $\angle CBD$ is larger than $\angle CBE$. Subtraction is used when the angle starts big and ends smaller. Addition is used when the angle starts small and ends bigger.

## EXERCISES

### EXERCISE 1–1

*Use the accompanying diagrams to answer each of the following questions.*

1. $\overline{BD}$ bisects $\angle ABC$. If $m\angle ABC = 60°$, find $m\angle ABD$.

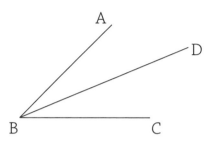

2. $\overline{AB}$ bisects $\overline{CD}$ at $E$. If $CE = 10$, find the measure of $CD$.

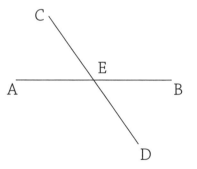

3. $\overline{AB}$ bisects $\overline{CD}$ at $E$. $CE = 3x + 5$, $ED = 7x - 15$. Find the length of $CD$.

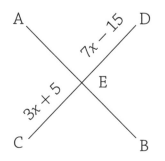

4. $\overline{AB}$ and $\overline{CD}$ intersect at $E$. $m\angle CEB = 2x$, $m\angle AED = x + 15$. Solve for $x$.

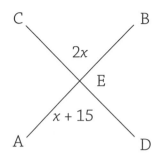

5. $\overline{AB} \perp \overline{CD}$ at $E$, $m\angle AED = 2x + 10$. Solve for $x$.

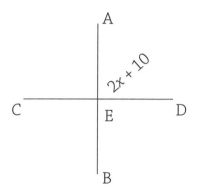

## EXERCISE 1–2

*Answer the following questions involving angles.*

1. $\angle A$ and $\angle B$ are complementary angles. If $m\angle A = 25°$, find $m\angle B$.

2. $\angle C$ and $\angle D$ are supplementary angles. If $m\angle C = 95°$, find $m\angle D$.

## EXERCISE 1–3

*Use the accompanying diagrams to answer each of the following questions.*

1. In the diagram of $\overline{ABD}$, $m\angle ABC = 12x + 5$ and $\angle mCBD = 8x - 25$. Solve for $x$.

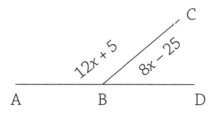

2. $A$ is the midpoint of $\overline{CT}$, $CA = 2x - 4$, $CT = 3x + 2$. Solve for $x$.

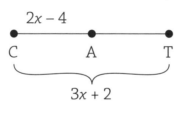

3. $\overline{GE}$ bisects $\angle DEF$, $m\angle DEG = 3x + 2$, $m\angle FEG = 5x - 10$. Find $m\angle DEG$.

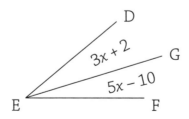

4. $\overline{ST}$ is the perpendicular bisector of $\overline{MP}$. Name a pair of segments and a pair of angles that you can conclude are congruent to one another.

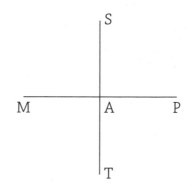

5. Use the diagram of $\overline{DRAW}$ to name the property that states $\overline{RA} \cong \overline{RA}$.

6. Find the sum of the two angles in the accompanying diagram.
$\angle PQS + \angle SQR = $ _____.

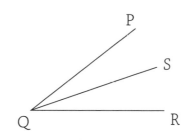

7. In the given diagram, $\overline{SR}$ represents a line segment and $\overline{SA} \cong \overline{RO}$.
   Fill in the missing segment that makes the equation true:
   $\overline{SA} - \overline{OA} \cong$ _____ $-\overline{OA}$.

## EXERCISE 1–4

*Write a formal proof using the following information:*

Given: $\overline{ABLE}$, $\overline{AB} \cong \overline{EL}$

Prove: $\overline{AL} \cong \overline{EB}$

# Triangle Proofs

## MUST KNOW

⚡ Geometric proofs provide reasons for factual statements using organized and logical explanations.

⚡ The five postulates that enable us to prove triangles are congruent are *side-side-side, side-angle-side, angle-side-angle, angle-angle-side,* and *hypotenuse-leg.*

⚡ If two triangles are congruent, their corresponding sides and angles are also congruent.

hat makes two triangles congruent? When two triangles are congruent, you can slide one triangle to fit exactly on top of the other triangle. The angles that match up with one another are called **corresponding angles**, and the sides that match up with one another are called **corresponding sides**. When two triangles are congruent, their corresponding angles are congruent. We represent this by placing a loop in the pair of corresponding angles. Sometimes we will place loops with slashes in them to differentiate which pairs of angles are congruent. When two triangles are congruent, their corresponding sides are congruent. We represent this by placing a slash in the pair of corresponding sides. We change the number of slashes we place in the sides to differentiate which pairs of sides are congruent. This is shown in the accompanying diagram of congruent triangles $\triangle ABC$ and $\triangle A'\,B'\,C'$.

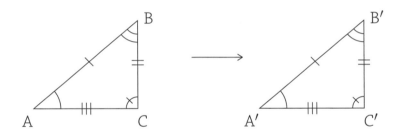

As you can see in the diagram, the corresponding congruent angles are $\angle A \cong \angle A'$, $\angle B \cong \angle B'$, and $\angle C \cong \angle C'$, and the corresponding congruent sides are $\overline{AB} \cong \overline{A'B'}$, $\overline{BC} \cong \overline{B'C'}$, and $\overline{AC} \cong \overline{A'C'}$.

The following example practices naming corresponding sides and angles of congruent triangles.

**BTW**

When naming two congruent triangles, the order of the letters will indicate the angles that correspond to each other. The order in which you name the triangles matters. If $\triangle RST \cong \triangle WXY$, then $\angle R \cong \angle W$, $\angle S \cong \angle X$, and $\angle T \cong \angle Y$.

▶ The accompanying diagram shows $\triangle JLU \cong \triangle GBT$. Name the three pairs of corresponding sides and three pairs of corresponding angles that are congruent.

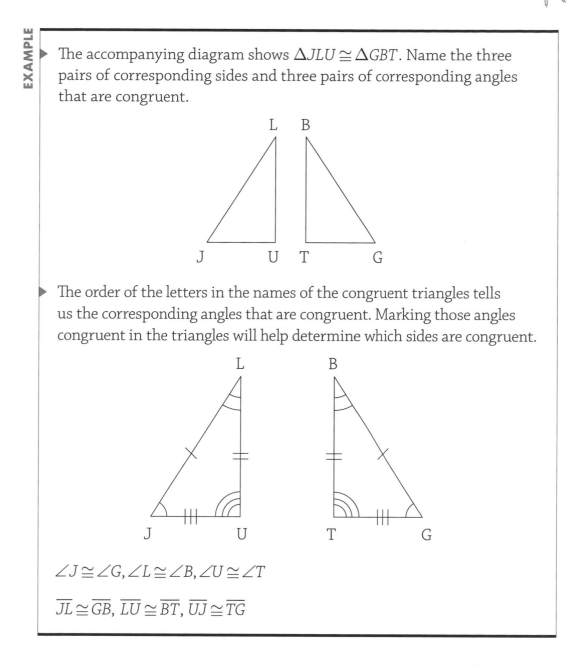

▶ The order of the letters in the names of the congruent triangles tells us the corresponding angles that are congruent. Marking those angles congruent in the triangles will help determine which sides are congruent.

$\angle J \cong \angle G, \angle L \cong \angle B, \angle U \cong \angle T$

$\overline{JL} \cong \overline{GB}, \overline{LU} \cong \overline{BT}, \overline{UJ} \cong \overline{TG}$

Two triangles are congruent if all six parts of one triangle are congruent to all six parts of the other triangle. There are postulates that allow us to prove triangles congruent by only showing three pairs of corresponding parts congruent.

## Side-Side-Side Postulate
## for Proving Triangles Congruent

The first method for proving triangles congruent is the **side-side-side (SSS)** postulate. If three sides of one triangle are congruent to three sides of another triangle, then the triangles are congruent. The following diagram shows $\triangle ABC$ and $\triangle DEF$ congruent to one another using the side-side-side postulate.

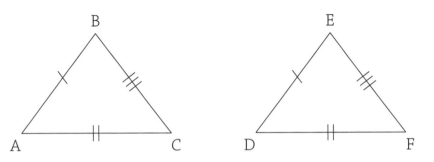

Can you name the two corresponding parts of the following triangles that would need to be congruent to prove the triangles congruent using the side-side-side postulate?

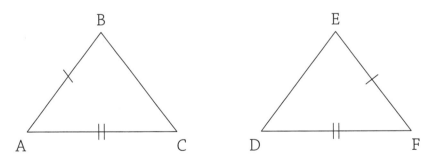

We already have two pairs of corresponding sides marked congruent, which means we need a third pair of corresponding sides congruent. Therefore, we would need $\overline{BC} \cong \overline{ED}$.

In the following example, we will use the side-side-side postulate in a formal geometric proof. Remember that in a formal geometric proof we use the information given in the problem to develop statements of sides or angles of the triangles that are congruent to one another. We support these statements by stating the reason for why each statement is true. That is why we set up a statement reason table for a formal geometric proof.

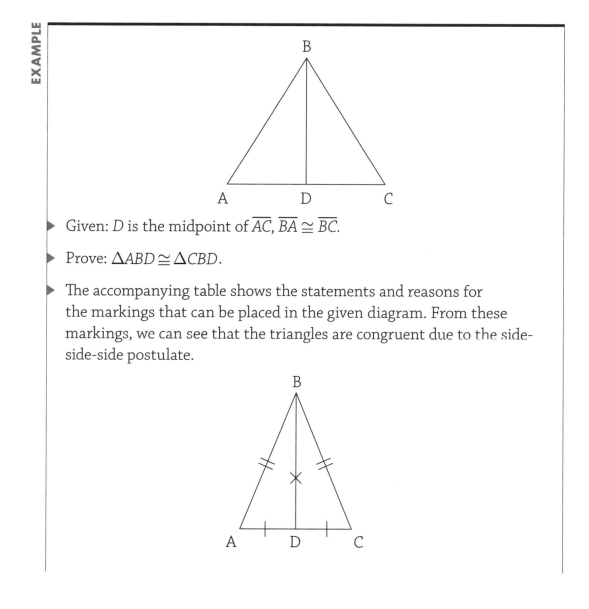

▶ Given: $D$ is the midpoint of $\overline{AC}$, $\overline{BA} \cong \overline{BC}$.

▶ Prove: $\triangle ABD \cong \triangle CBD$.

▶ The accompanying table shows the statements and reasons for the markings that can be placed in the given diagram. From these markings, we can see that the triangles are congruent due to the side-side-side postulate.

| Statement | Reason |
|---|---|
| 1. $D$ is the midpoint of $\overline{AC}$. | 1. Given |
| 2. $\overline{AD} \cong \overline{CD}$ | 2. A midpoint divides a segment into two congruent segments. |
| 3. $\overline{BA} \cong \overline{BC}$ | 3. Given |
| 4. $\overline{BD} \cong \overline{BD}$ | 4. Reflexive property |
| 5. $\triangle ABD \cong \triangle CBD$ | 5. SSS $\cong$ SSS |

## Side-Angle-Side Postulate for Proving Triangles Congruent

The second method for proving triangles congruent is the **side-angle-side (SAS)** postulate. If two sides and the included angle of one triangle are congruent to two sides and the included angle of another triangle, then the triangles are congruent.

The included angle is referred to as the angle that is between the two sides of a triangle. The following diagram shows $\triangle ABC$ and $\triangle DEF$ congruent to one another using the side-angle-side postulate. Notice that the angle that is marked is between the two sides that are marked.

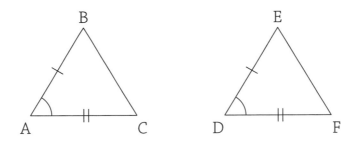

Can you name the two corresponding parts of the following triangles that would need to be congruent to prove the triangles congruent using the side-angle-side postulate?

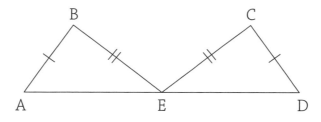

We already have two pairs of corresponding sides marked congruent, which means we need the included pair of corresponding angles congruent. This means the angles must be located between the two congruent sides. Therefore, we would need $\angle B \cong \angle C$.

Let's use the side-angle-side postulate in a formal geometric proof.

**EXAMPLE**

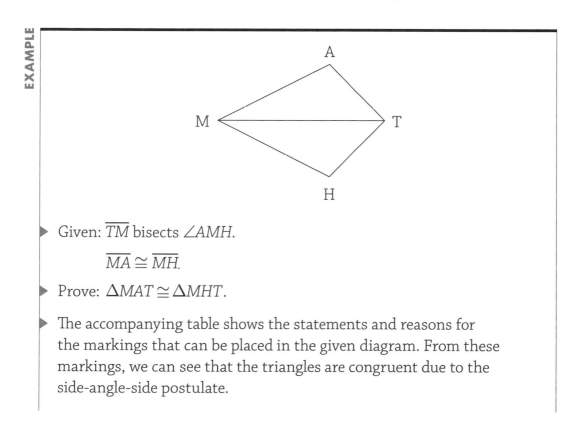

▶ Given: $\overline{TM}$ bisects $\angle AMH$.

      $\overline{MA} \cong \overline{MH}$.

▶ Prove: $\triangle MAT \cong \triangle MHT$.

▶ The accompanying table shows the statements and reasons for the markings that can be placed in the given diagram. From these markings, we can see that the triangles are congruent due to the side-angle-side postulate.

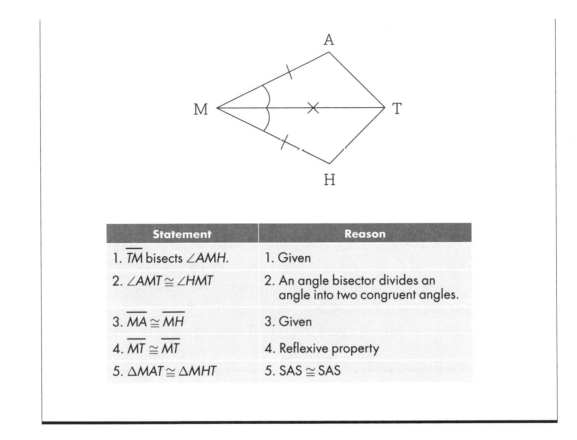

| Statement | Reason |
|---|---|
| 1. $\overline{TM}$ bisects $\angle AMH$. | 1. Given |
| 2. $\angle AMT \cong \angle HMT$ | 2. An angle bisector divides an angle into two congruent angles. |
| 3. $\overline{MA} \cong \overline{MH}$ | 3. Given |
| 4. $\overline{MT} \cong \overline{MT}$ | 4. Reflexive property |
| 5. $\triangle MAT \cong \triangle MHT$ | 5. SAS $\cong$ SAS |

## Angle-Side-Angle Postulate for Proving Triangles Congruent

The third method for proving triangles congruent is the **angle-side-angle (ASA)** postulate. If two angles and the included side of one triangle are congruent to two angles and the included side of another triangle, then the triangles are congruent. The following diagram shows $\triangle ABC$ and $\triangle DEF$ congruent to one another using the angle-side-angle postulate. Notice that the side that is marked is between the two angles that are marked.

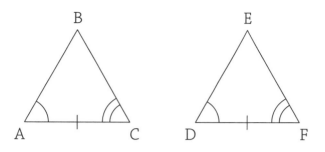

Can you name the two corresponding parts of the following triangles that would need to be congruent to prove the triangles congruent using the angle-side-angle postulate?

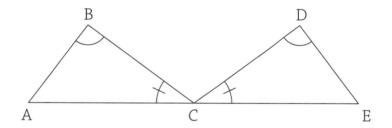

We already have two pairs of corresponding angles marked congruent, which means we need the included pair of corresponding sides congruent. This means the sides must be located between the two congruent angles. Therefore, we would need $\overline{BC} \cong \overline{DC}$.

In the following example, we will use the angle-side-angle postulate in a formal geometric proof.

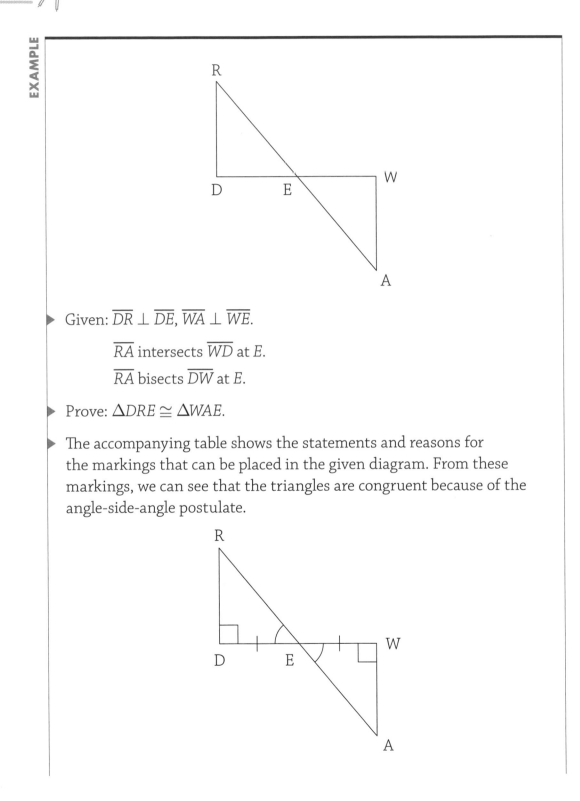

▶ Given: $\overline{DR} \perp \overline{DE}$, $\overline{WA} \perp \overline{WE}$.

   $\overline{RA}$ intersects $\overline{WD}$ at $E$.

   $\overline{RA}$ bisects $\overline{DW}$ at $E$.

▶ Prove: $\triangle DRE \cong \triangle WAE$.

▶ The accompanying table shows the statements and reasons for the markings that can be placed in the given diagram. From these markings, we can see that the triangles are congruent because of the angle-side-angle postulate.

| Statement | Reason |
|---|---|
| 1. $\overline{DR} \perp \overline{DE}$, $\overline{WA} \perp \overline{WE}$ | 1. Given |
| 2. $\angle RDE$ and $\angle AWE$ are right angles. | 2. Perpendicular lines form right angles. |
| 3. $\angle RDE \cong \angle AWE$ | 3. Right angles are congruent. |
| 4. $\overline{RA}$ intersects $\overline{WD}$ at $E$. | 4. Given |
| 5. $\angle DER$ and $\angle WEA$ are vertical angles. | 5. Intersecting lines form vertical angles. |
| 6. $\angle DER \cong \angle WEA$ | 6. Vertical angles are congruent. |
| 7. $\overline{RA}$ bisects $\overline{DW}$ at $E$. | 7. Given |
| 8. $\overline{DE} \cong \overline{WE}$ | 8. A bisector divides a segment into two congruent segments. |
| 9. $\triangle DRE \cong \triangle WAE$ | 9. ASA $\cong$ ASA |

## Angle-Angle-Side Postulate for Proving Triangles Congruent

The fourth method for proving triangles congruent is the **angle-angle-side (AAS)** postulate. If two angles and a non-included side of one triangle are congruent to the two angles and a non-included side of another triangle, then the triangles are congruent.

Be careful! A common mistake is to confuse AAS with ASA. If the marked side is between the marked angles, you would use ASA; but if the marked side is not between the marked angles, you would use AAS. The following diagram shows $\triangle ABC$ and $\triangle DEF$ congruent to one another using the angle-angle-side postulate. Notice that the side that is marked is not between the two angles that are marked.

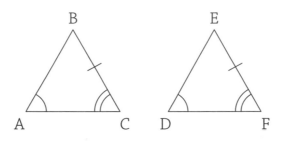

Can you name the two corresponding parts of the following triangles that would need to be congruent to prove the triangles congruent using the angle-angle-side postulate?

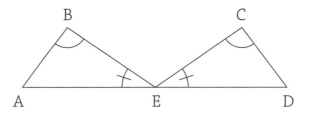

We already have two pairs of corresponding angles marked congruent, which means we need the pair of corresponding sides that are *not included* to be congruent. This means the sides cannot be located between the two congruent angles. Therefore, we can either state $\overline{BA} \cong \overline{CD}$ or $\overline{AE} \cong \overline{DE}$.

Let's use the angle-angle-side postulate in a formal geometric proof.

EXAMPLE

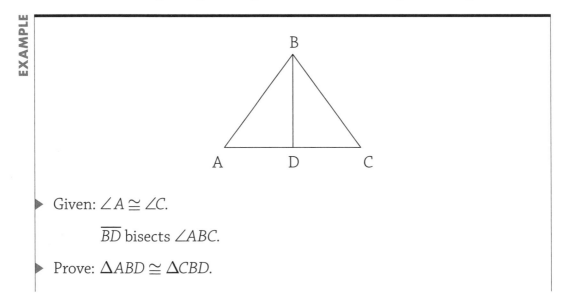

▶ Given: $\angle A \cong \angle C$.

$\overline{BD}$ bisects $\angle ABC$.

▶ Prove: $\triangle ABD \cong \triangle CBD$.

▶ The accompanying table shows the statements and reasons for the markings that can be placed in the given diagram. From these markings, we can see that the triangles are congruent because of the angle-angle-side postulate.

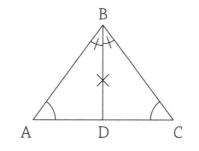

| Statement | Reason |
|---|---|
| 1. $\angle A \cong \angle C$ | 1. Given |
| 2. $\overline{BD}$ bisects $\angle ABC$. | 2. Given |
| 3. $\angle ABD \cong \angle CBD$ | 3. A bisector divides an angle into two congruent angles. |
| 4. $\overline{BD} \cong \overline{BD}$ | 4. Reflexive property |
| 5. $\triangle ABD \cong \triangle CBD$ | 5. AAS $\cong$ AAS |

## Why Is Side-Side-Angle Not a Postulate for Proving Triangles Congruent?

The following illustration of isosceles triangle *ABC* shows why SSA or ASS is *not* a method that can be used to prove triangles congruent.

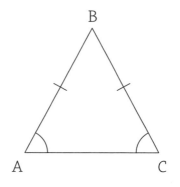

You can see in this diagram that $\overline{AB} \cong \overline{CB}$ and $\angle A \cong \angle C$. Let's draw $\overline{BD}$ in such a way that $D$ is not the midpoint of $\overline{AC}$. It creates two triangles that we can clearly see are *not congruent*, even though we could use the reflexive property with $\overline{BD}$ and create markings that show side-side-angle.

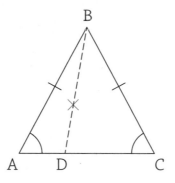

## Why Is Angle-Angle-Angle Not a Postulate for Proving Triangles Congruent?

The following illustration of two equilateral triangles shows why triangles cannot be proved congruent by having three pairs of corresponding angles congruent.

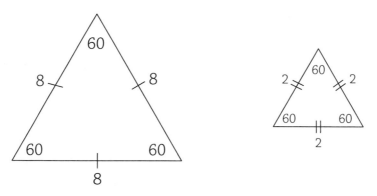

You can see that both triangles have three pairs of corresponding angles congruent because they are all equal to 60°, but they are clearly *not congruent* because one equilateral triangle has sides measuring 8 and the other has sides measuring 2.

# Hypotenuse-Leg Postulate
# for Proving Triangles Congruent

The last method for proving triangles congruent is the **hypotenuse-leg (HL) postulate**. If the hypotenuse and leg of a right triangle are congruent to the hypotenuse and leg of another right triangle, then the right triangles are congruent. Always remember that this postulate can only be used for right triangles. This means that to use this postulate we first must have right angles and therefore be able to say that the triangles are right triangles. The following diagram shows $\triangle ABC$ and $\triangle DEF$ congruent to one another using the hypotenuse-leg postulate.

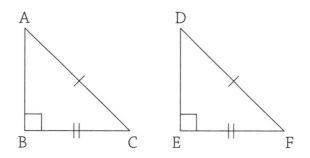

Can you name the two corresponding parts of the following triangles that would need to be congruent to prove the triangles congruent using the hypotenuse-leg postulate?

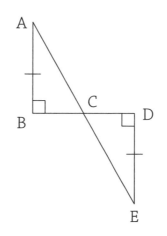

We already have a pair of right angles marked congruent. This shows that our triangles are right triangles. We also have a pair of corresponding legs of the right triangles congruent. This means we need to have the hypotenuse of both triangles marked congruent. Because the hypotenuse of a triangle is found opposite the right angle, we need $\overline{AC} \cong \overline{EC}$.

In the following example, we will use the hypotenuse-leg postulate in a formal geometric proof.

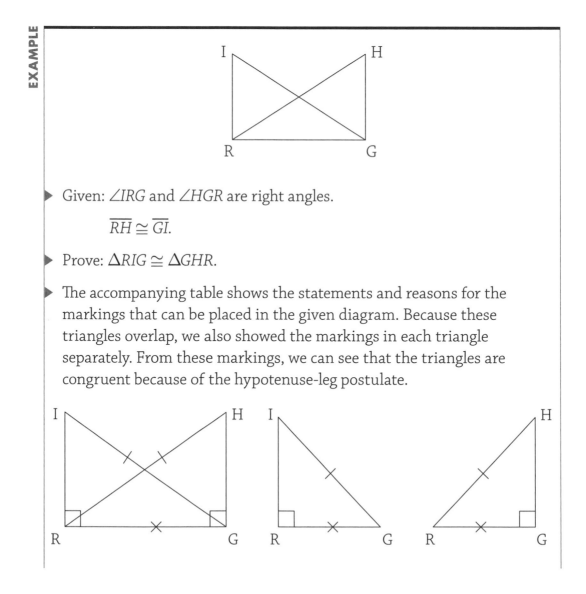

▶ Given: $\angle IRG$ and $\angle HGR$ are right angles.

   $\overline{RH} \cong \overline{GI}$.

▶ Prove: $\triangle RIG \cong \triangle GHR$.

▶ The accompanying table shows the statements and reasons for the markings that can be placed in the given diagram. Because these triangles overlap, we also showed the markings in each triangle separately. From these markings, we can see that the triangles are congruent because of the hypotenuse-leg postulate.

| Statement | Reason |
|---|---|
| 1. ∠*IRG* and ∠*HGR* are right angles. | 1. Given |
| 2. △*IRG* and △*HGR* are right triangles. | 2. If a triangle has a right angle, then it is a right triangle. |
| 3. $\overline{RH} \cong \overline{GI}$ | 3. Given |
| 4. $\overline{RG} \cong \overline{RG}$ | 4. Reflexive property |
| 5. △*RIG* ≅ △*GHR* | 5. HL ≅ HL |

▶ Right triangles can be drawn in all different sizes. Remember that just because two triangles are right triangles that does not mean they are congruent. Once we show that we have right triangles, we still need to show that the hypotenuse and leg of one of the right triangles are congruent to the hypotenuse and leg of the other right triangle.

Now that we have learned the five different postulates for triangle congruence, we will apply them to geometric proofs using geometric terms, such as isosceles triangles, and altitudes of a triangle. We will also work with more overlapping triangles.

The following example requires you to prove triangles congruent when given information about an isosceles triangle.

**EXAMPLE**

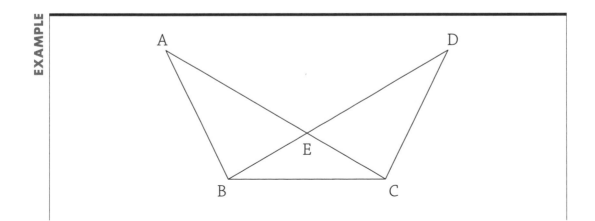

▸ Given: △BEC is isosceles with vertex E.

$\overline{AC}$ intersects $\overline{DB}$ at E.

∠A ≅ ∠D.

▸ Prove: △ABE ≅ △DCE.

▸ The accompanying table shows the statements and reasons for the markings that can be placed in the given diagram. From these markings, we can see that the triangles are congruent because of the angle-angle-side postulate.

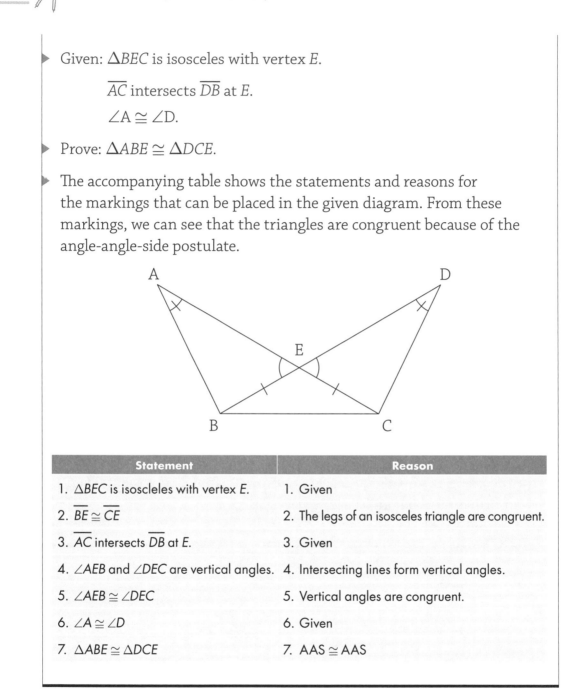

| Statement | Reason |
|---|---|
| 1. △BEC is isoscleles with vertex E. | 1. Given |
| 2. $\overline{BE} \cong \overline{CE}$ | 2. The legs of an isosceles triangle are congruent. |
| 3. $\overline{AC}$ intersects $\overline{DB}$ at E. | 3. Given |
| 4. ∠AEB and ∠DEC are vertical angles. | 4. Intersecting lines form vertical angles. |
| 5. ∠AEB ≅ ∠DEC | 5. Vertical angles are congruent. |
| 6. ∠A ≅ ∠D | 6. Given |
| 7. △ABE ≅ △DCE | 7. AAS ≅ AAS |

The next example uses the same diagram but requires you to prove the two overlapping triangles are congruent. The set of given information is slightly different, which will lead to a completely different set of steps in the proof.

EXAMPLE

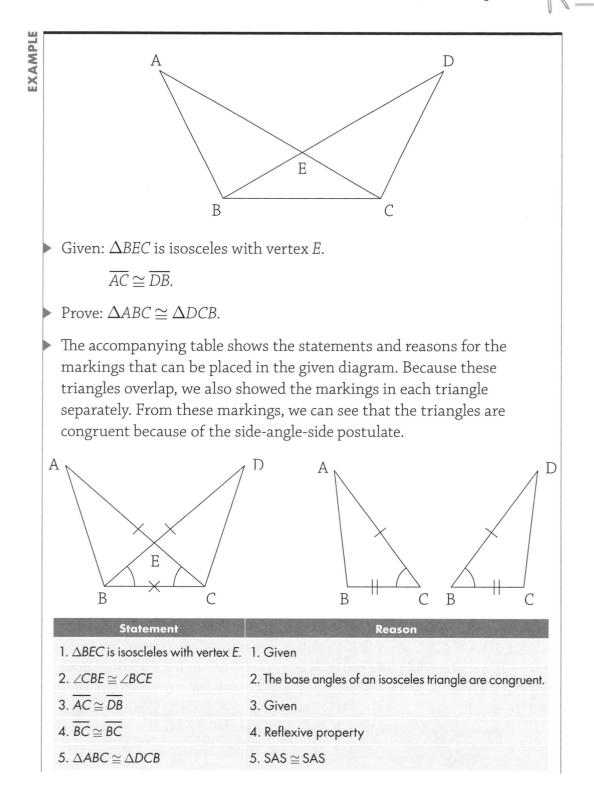

▶ Given: $\triangle BEC$ is isosceles with vertex $E$.

   $\overline{AC} \cong \overline{DB}$.

▶ Prove: $\triangle ABC \cong \triangle DCB$.

▶ The accompanying table shows the statements and reasons for the markings that can be placed in the given diagram. Because these triangles overlap, we also showed the markings in each triangle separately. From these markings, we can see that the triangles are congruent because of the side-angle-side postulate.

| Statement | Reason |
|---|---|
| 1. $\triangle BEC$ is isoscleles with vertex $E$. | 1. Given |
| 2. $\angle CBE \cong \angle BCE$ | 2. The base angles of an isosceles triangle are congruent. |
| 3. $\overline{AC} \cong \overline{DB}$ | 3. Given |
| 4. $\overline{BC} \cong \overline{BC}$ | 4. Reflexive property |
| 5. $\triangle ABC \cong \triangle DCB$ | 5. SAS $\cong$ SAS |

▶ Did you see the difference in step 2 between the two proofs? In this proof we needed to use the fact that the base angles of an isosceles triangle rather than the legs are congruent. Overlapping triangles can be challenging to work with. We recommend that you use a highlighter to tell the triangles apart or slide the triangles apart and view them separately as we showed in the diagram.

---

## EXTRA HELP

**Triangle Congruence Theorems**

**Only use these:**

$SSS \cong SSS$

$SAS \cong SAS$

$ASA \cong ASA$

$AAS \cong AAS$

$HL \cong HL$

**Never use these:**

$SSA \cong SSA$

$AAA \cong AAA$

---

# Corresponding Parts of Congruent Triangles Are Congruent

When two triangles are congruent, the angles and sides of one of the triangles are congruent to the corresponding angles and sides of the other triangle. We use the term **corresponding parts of congruent triangles are congruent** (CPCTC) for this theorem. We use this theorem at the end of a proof when we have shown that two triangles are congruent but are looking to prove that a pair of corresponding sides or angles are congruent.

The following example requires the use of the CPCTC Theorem to prove two sides congruent. It also incorporates the term *altitude*, which is a segment that extends from any angle of a triangle and is perpendicular to a line containing the opposite side.

> **BTW**
>
> *Before using CPCTC, you must prove that the triangles are congruent first.*

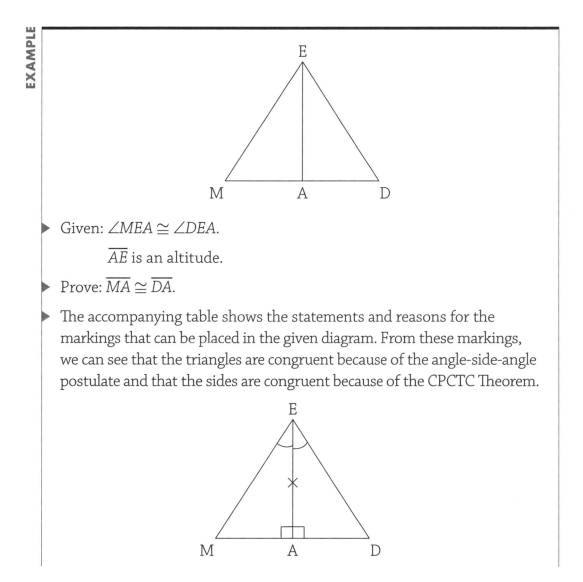

▶ Given: $\angle MEA \cong \angle DEA$.

$\overline{AE}$ is an altitude.

▶ Prove: $\overline{MA} \cong \overline{DA}$.

▶ The accompanying table shows the statements and reasons for the markings that can be placed in the given diagram. From these markings, we can see that the triangles are congruent because of the angle-side-angle postulate and that the sides are congruent because of the CPCTC Theorem.

| Statement | Reason |
|---|---|
| 1. $\angle MEA \cong \angle DEA$ | 1. Given |
| 2. $\overline{AE}$ is an altitude. | 2. Given |
| 3. $\angle MAE$ and $\angle DAE$ are right angles. | 3. Altitudes form right angles. |
| 4. $\angle MAE \cong \angle DAE$ | 4. Right angles are congruent. |
| 5. $\overline{AE} \cong \overline{AE}$ | 5. Reflexive property |
| 6. $\triangle MEA \cong \triangle DEA$ | 6. ASA $\cong$ ASA |
| 7. $\overline{MA} \cong \overline{DA}$ | 7. CPCTC |

The following example requires the use of the CPCTC Theorem to prove two angles congruent. It also incorporates the subtraction postulate that we reviewed in the previous chapter.

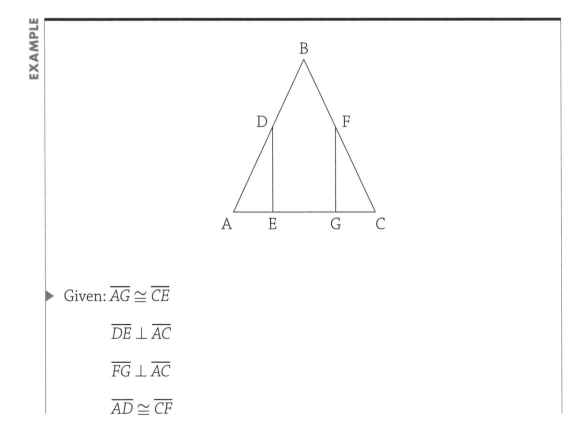

Given: $\overline{AG} \cong \overline{CE}$

$\overline{DE} \perp \overline{AC}$

$\overline{FG} \perp \overline{AC}$

$\overline{AD} \cong \overline{CF}$

▶ Prove: $\angle ADE \cong \angle CFG$.

▶ The accompanying table shows the statements and reasons for the markings that can be placed in the given diagram. From these markings, we can see that the triangles are congruent because of the hypotenuse-leg postulate and that the angles are congruent because of the CPCTC Theorem.

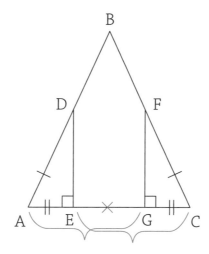

| Statement | Reason |
|---|---|
| 1. $\overline{AG} \cong \overline{CE}$ | 1. Given |
| 2. $\overline{EG} \cong \overline{EG}$ | 2. Reflexive property |
| 3. $\overline{AE} \cong \overline{CG}$ | 3. Subtraction postulate |
| 4. $\overline{DE} \perp \overline{AC}, \overline{FG} \perp \overline{AC}$ | 4. Given |
| 5. $\angle AED$ and $\angle CGF$ are right angles. | 5. Perpendicular lines form right angles. |
| 6. $\angle AED \cong \angle CGF$ | 6. Right angles are congruent. |
| 7. $\triangle AED$ and $\triangle CGF$ are right triangles. | 7. If a triangle has a right angle, then it is a right triangle. |
| 8. $\overline{AD} \cong \overline{CF}$ | 8. Given |
| 9. $\triangle AED \cong \triangle CGF$ | 9. HL $\cong$ HL |
| 10. $\angle ADE \cong \angle CFG$ | 10. CPCTC |

## EXERCISES

### EXERCISE 2-1

*Name the postulate that would prove each the following pairs of triangles congruent.*

1.

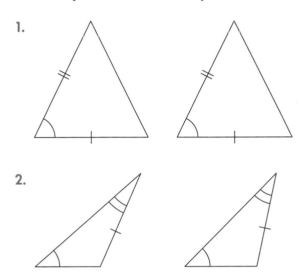

2.

3.

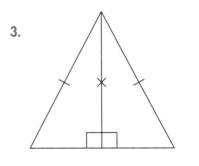

4.

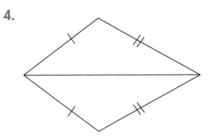

## EXERCISE 2-2

*Use the given accompanying diagrams to answer each question.*

1. Which pair of corresponding sides is needed to prove $\triangle RST \cong \triangle XWT$ in the accompanying diagram using ASA?

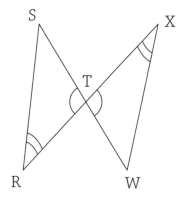

2. Which pair of corresponding sides is needed to prove $\triangle RST \cong \triangle XWT$ in the accompanying diagram using AAS?

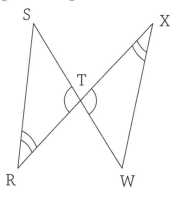

## EXERCISE 2-3

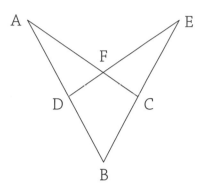

*Fill in the missing information for the following proof.*

Given: $\angle A \cong \angle E$

$\overline{AD} \cong \overline{EC}$

$\overline{BD} \cong \overline{BC}.$

Prove: $\triangle ABC \cong \triangle EBD.$

| Statement | Reason |
|---|---|
| $\angle A \cong \angle E$ | Given |
| $\overline{AD} \cong \overline{EC}, \overline{BD} \cong \overline{BC}$ | 1. |
| 2. | Addition postulate |
| 3. | Reflexive property |
| $\triangle ABC \cong \triangle EBD$ | 4. |

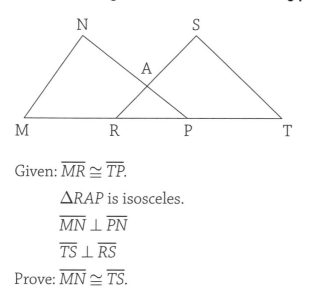

## EXERCISE 2-4

*Fill in the missing information for the following proof.*

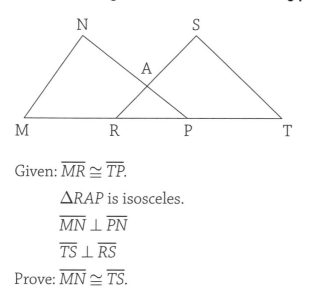

Given: $\overline{MR} \cong \overline{TP}$.

$\Delta RAP$ is isosceles.

$\overline{MN} \perp \overline{PN}$

$\overline{TS} \perp \overline{RS}$

Prove: $\overline{MN} \cong \overline{TS}$.

| Statement | Reason |
|---|---|
| $\overline{MR} \cong \overline{TP}$ | Given |
| $\overline{RP} \cong \overline{RP}$ | 1. |
| $\overline{MP} \cong \overline{TR}$ | 2. |
| $\Delta RAP$ is isosceles. | Given |
| 3. | If a triangle is isosceles, then its base angles are congruent. |
| $\overline{MN} \perp \overline{PN}$<br>$\overline{TS} \perp \overline{RS}$ | Given |
| $\angle N$ and $\angle S$ are right angles. | Perpendicular lines form right angles. |
| $\angle N \cong \angle S$ | Right angles are congruent. |
| 4. | AAS ≅ AAS |
| $\overline{MN} \cong \overline{TS}$ | 5. |

## EXERCISE 2-5

*Use the accompanying diagrams to answer each question.*

1. $\triangle ABC \cong \triangle RST$, $m\angle C = 2x + 40$, and $m\angle T = 8x - 20$. Solve for x.

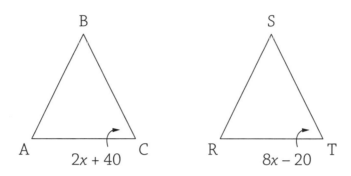

2. $\triangle ABC \cong \triangle RST$. If $AC = 14x - 4$ and $RT = 2x + 20$, solve for x.

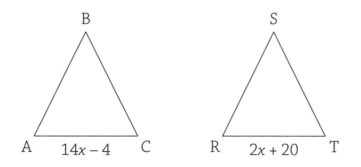

## EXERCISE 2-6

*Complete each of the following geometric proofs.*

1. Given: $\overline{PN}$ and $\overline{MQ}$ bisect each other at $E$.

   Prove: $\angle M \cong \angle Q$.

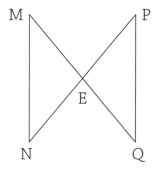

2. Given: $\overline{BD}$ bisects $\angle HBT$.

   $\angle HDB \cong \angle TDB$.

   Prove: $\triangle BHD \cong \triangle BTD$.

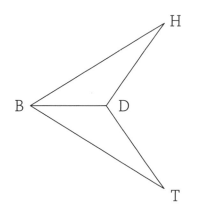

3. Given: $\overline{AE} \cong \overline{CD}$.

   $\triangle ABC$ is isosceles with vertex $B$.

   Prove: $\triangle ABD \cong \triangle CBE$.

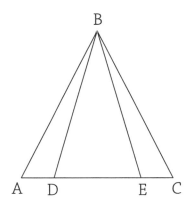

Flashcard App

# 3 Classifying Triangles

## MUST ⚡ KNOW

 The angle measurements and side lengths of a triangle determine whether it is *acute*, *obtuse*, *isosceles*, *equilateral*, *scalene*, or *right*.

 Knowing the properties of the different kinds of triangles allows us to solve for unknown sides and angles algebraically.

⚡ The sum of the angles of a triangle equals 180°.

⚡ The largest side of a triangle is always found opposite the largest angle of the triangle.

n this chapter we will focus on the basic properties of triangles and also discuss the properties for specific types of triangles. The classification of triangles with sides and angles provides information that will help you with both algebraic style problems and many different geometric proofs.

## Solving for the Angles in a Triangle

A **triangle** is a **polygon** that contains three sides and three angles called **vertices**. One of the most important properties of a triangle is that the sum of the three angles of a triangle must equal $180°$. This property gives us the ability to use algebra to solve for angle measurements, as seen in the following example.

**EXAMPLE**

▶ In the accompanying diagram of $\triangle DOG$, $m\angle D = 3x + 2$, $m\angle O = 2x + 5$, and $m\angle G = 6x + 8$. Solve for $x$ and then find the measure of each angle of the triangle.

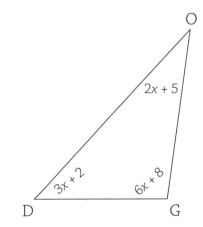

▶ We can set the sum of the three angles of the triangle equal to 180° to solve for $x$ and then substitute the $x$ value into the angle expressions to find the measure of each angle.

$$3x + 2 + 6x + 8 + 2x + 5 = 180$$

$$11x + 15 = 180$$

$$\begin{array}{r} -15 \quad -15 \\ \hline \dfrac{11x}{11} = \dfrac{165}{11} \end{array}$$

$$x = 15$$

$$m\angle D = 3(15) + 2 = 47°$$

$$m\angle O = 2(15) + 5 = 35°$$

$$m\angle G = 6(15) + 8 = 98°$$

# Exterior Angle Theorem

The **exterior angle** of a triangle is formed when we extend a side of a triangle through a vertex of the triangle. The exterior angle is adjacent to the angle inside the triangle. The accompanying diagram shows $\triangle CAT$ where $\overline{CT}$ is extended through $T$ to $M$, which makes $\angle ATM$ an exterior angle.

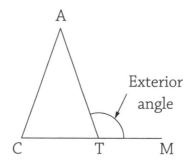

The **exterior angle theorem** says that the exterior angle of a triangle is equal to the sum of the two nonadjacent interior angles of the triangle. This theorem is especially helpful in finding the measures of the angles in a triangle. The following diagram shows an exterior angle with the nonadjacent interior angles labeled.

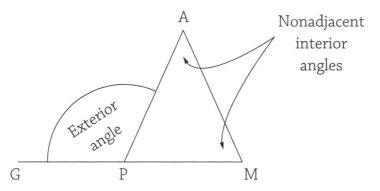

We will prove the exterior angle theorem using the following diagram, which shows $\triangle DEF$ where $\overline{DF}$ is extended through $F$ to $G$ to form an exterior angle at $\angle EFG$ and nonadjacent interior angles at $\angle D$ and $\angle E$.

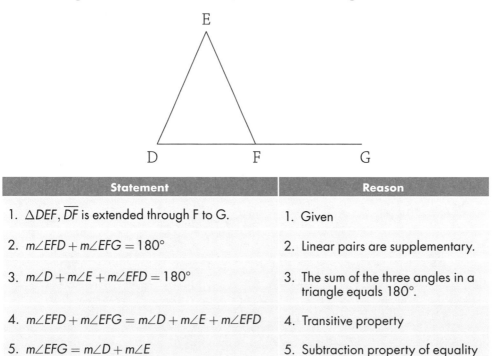

| Statement | Reason |
|---|---|
| 1. $\triangle DEF$, $\overline{DF}$ is extended through F to G. | 1. Given |
| 2. $m\angle EFD + m\angle EFG = 180°$ | 2. Linear pairs are supplementary. |
| 3. $m\angle D + m\angle E + m\angle EFD = 180°$ | 3. The sum of the three angles in a triangle equals 180°. |
| 4. $m\angle EFD + m\angle EFG = m\angle D + m\angle E + m\angle EFD$ | 4. Transitive property |
| 5. $m\angle EFG = m\angle D + m\angle E$ | 5. Subtraction property of equality |

The following example will apply the exterior angle theorem to a basic algebraic problem.

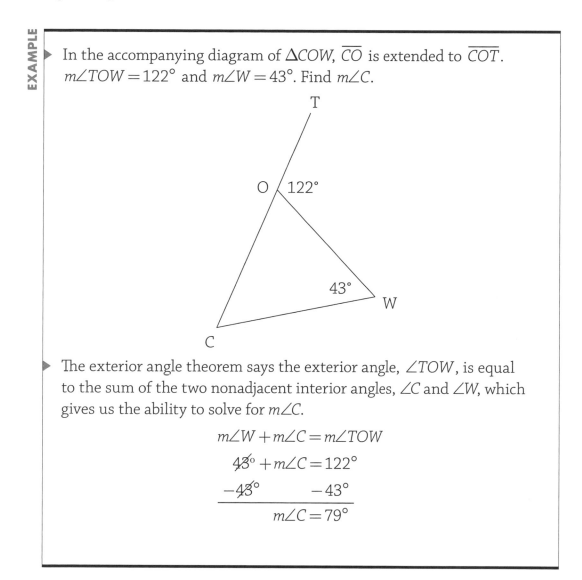

**EXAMPLE**

▶ In the accompanying diagram of $\triangle COW$, $\overline{CO}$ is extended to $\overline{COT}$. $m\angle TOW = 122°$ and $m\angle W = 43°$. Find $m\angle C$.

▶ The exterior angle theorem says the exterior angle, $\angle TOW$, is equal to the sum of the two nonadjacent interior angles, $\angle C$ and $\angle W$, which gives us the ability to solve for $m\angle C$.

$$m\angle W + m\angle C = m\angle TOW$$
$$43° + m\angle C = 122°$$
$$\underline{-43° \qquad\quad -43°}$$
$$m\angle C = 79°$$

The preceding example could have been solved without the exterior angle theorem by using the knowledge of linear pairs and the fact that the sum of the angles in a triangle equals 180°. Before moving on, we should look at an example that can only be solved by using the exterior angle theorem.

This situation would occur if we do not know any of the actual angle measurements in the triangle, as seen in the following example.

In the accompanying diagram of $\triangle RST$, $\overline{RT}$ is extended to $\overline{RTU}$, $m\angle STU = 6x - 4$, $m\angle S = 2x + 5$, and $m\angle R = 3x$. Solve for $x$.

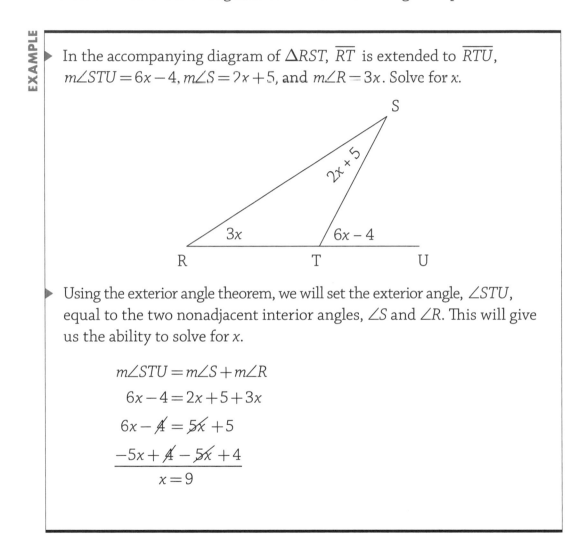

Using the exterior angle theorem, we will set the exterior angle, $\angle STU$, equal to the two nonadjacent interior angles, $\angle S$ and $\angle R$. This will give us the ability to solve for $x$.

$$m\angle STU = m\angle S + m\angle R$$
$$6x - 4 = 2x + 5 + 3x$$
$$6x - \cancel{4} = \cancel{5}x + 5$$
$$\underline{-5x + \cancel{4} - \cancel{5}x + 4}$$
$$x = 9$$

## Classifying Triangles by Angle Measurements

We can classify or name triangles based on the measurements of the angles within them. Let's begin with **acute triangles**. We know that an acute angle is an angle measuring between 0° and 90°. In an acute triangle, all three angles are acute angles. The following example shows a triangle that we can classify as an acute triangle.

In the accompanying diagram of $\triangle ABC$, $m\angle A = 5x + 12$, $m\angle B = 9x - 7$, and $m\angle C = 4x - 5$. Prove algebraically that this is an acute triangle.

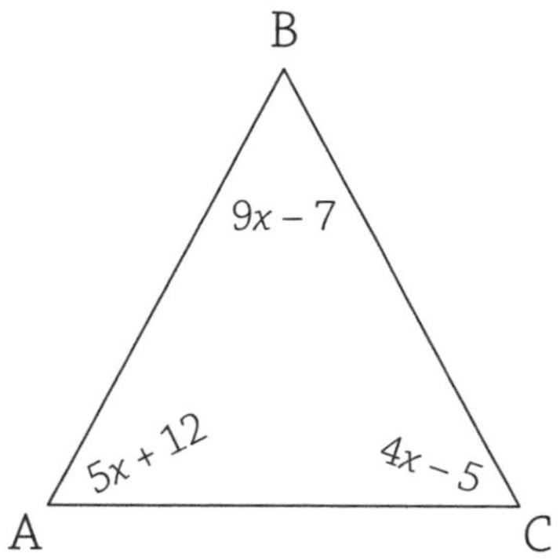

We can solve for $x$ using the knowledge that the sum of the three angles in a triangle equals 180°. We can then substitute the value of $x$ into the angle expressions to show that each angle is an angle between 0° and 90°, which will classify the triangle as an acute triangle.

$$m\angle A + m\angle B + m\angle C = 180°$$
$$5x + 12 + 9x - 7 + 4x - 5 = 180$$
$$\frac{\cancel{18}x}{\cancel{18}} = \frac{180}{18}$$
$$x = 10$$

$$m\angle A = 5x + 12 = 5(10) + 12 = 62°$$
$$m\angle B = 9x - 7 = 9(10) - 7 = 83°$$
$$m\angle C = 4x - 5 = 4(10) - 5 = 35°$$

All three angles of $\triangle ABC$ are acute, which makes $\triangle ABC$ an acute triangle.

The measures of the angles of a triangle can also classify a triangle as an **obtuse triangle**. An obtuse angle is an angle measuring between 90° and 180°. An obtuse triangle is a triangle that contains *one* obtuse angle.

Our second example shows a triangle that we can classify as obtuse based on its angle measures.

**BTW**

*A triangle cannot have more than one obtuse angle because the sum of the three angles would be larger than 180°.*

EXAMPLE

▶ In the accompanying diagram of $\triangle BAT$, $m\angle B = 3x - 11$, $m\angle A = 5x + 15$, and $m\angle T = 2x - 4$. Prove algebraically that this is an obtuse triangle.

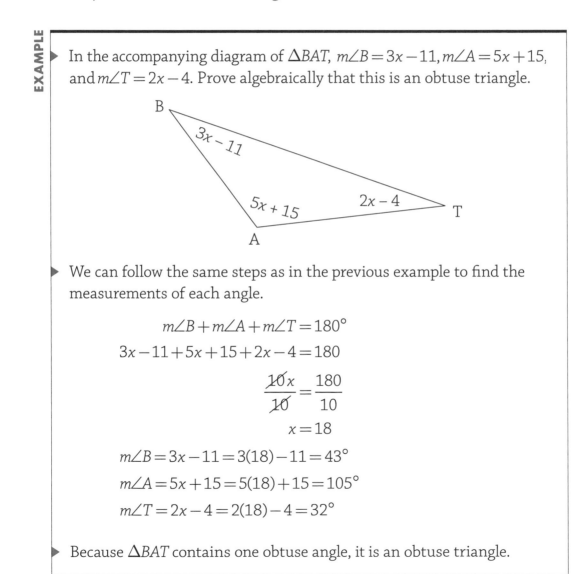

▶ We can follow the same steps as in the previous example to find the measurements of each angle.

$$m\angle B + m\angle A + m\angle T = 180°$$
$$3x - 11 + 5x + 15 + 2x - 4 = 180$$
$$\frac{\cancel{10}x}{\cancel{10}} = \frac{180}{10}$$
$$x = 18$$
$$m\angle B = 3x - 11 = 3(18) - 11 = 43°$$
$$m\angle A = 5x + 15 = 5(18) + 15 = 105°$$
$$m\angle T = 2x - 4 = 2(18) - 4 = 32°$$

▶ Because $\triangle BAT$ contains one obtuse angle, it is an obtuse triangle.

The next example will show that solving for the angles in a triangle can determine if a triangle is a **right triangle** by determining if one of the angles is a right angle, 90°. This chapter will not be focusing on classifying right triangles by their side lengths. We will be discussing that in the chapter dedicated to right triangles.

▶ In the following diagram of $\triangle BAM$, $m\angle B = 3x + 6$, $m\angle A = 7x + 6$, and $m\angle M = 4x$. Prove algebraically that this is a right triangle.

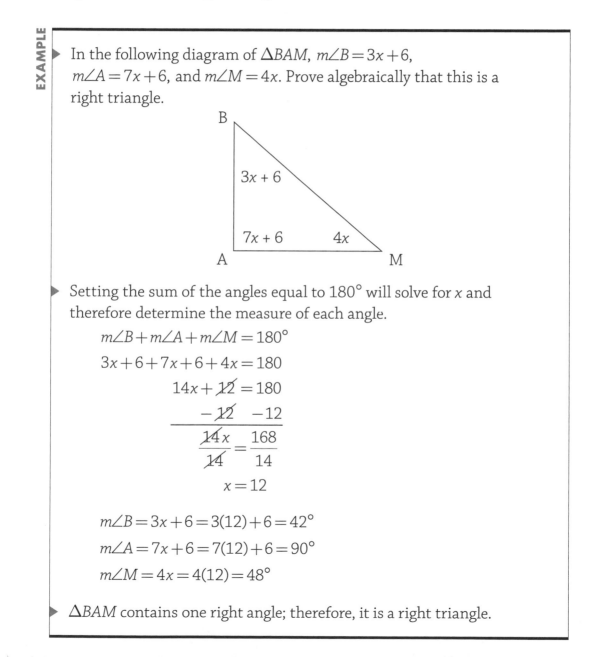

▶ Setting the sum of the angles equal to 180° will solve for $x$ and therefore determine the measure of each angle.

$$m\angle B + m\angle A + m\angle M = 180°$$
$$3x + 6 + 7x + 6 + 4x = 180$$
$$14x + 12 = 180$$
$$\underline{\phantom{14x} -12 \quad -12}$$
$$\frac{14x}{14} = \frac{168}{14}$$
$$x = 12$$

$$m\angle B = 3x + 6 = 3(12) + 6 = 42°$$
$$m\angle A = 7x + 6 = 7(12) + 6 = 90°$$
$$m\angle M = 4x = 4(12) = 48°$$

▶ $\triangle BAM$ contains one right angle; therefore, it is a right triangle.

## Isosceles, Equilateral, and Scalene Triangles

An isosceles triangle can be classified using both its angle measure and its side length. An isosceles triangle is a triangle that contains two equal sides. The angles opposite the two equal sides, the **base angles**, are also equal. Note that the angle formed from the two equal sides is called the **vertex angle**. This information is shown in the accompanying diagram of isosceles triangle $BLT$, where $\overline{BL} \cong \overline{LT}$, making $\angle B$ and $\angle T$ the congruent base angles and $\angle L$ the vertex angle.

**BTW**

*Most students like to draw their isosceles triangles with the vertex angle at the top and the base angles at the bottom. When you go to label your triangle, read the question carefully. The letter that should be at the top, the vertex, should be the letter that appears in both congruent sides. For example, if $\overline{AB} \cong \overline{BC}$, $\angle B$ is the vertex angle because B is in both congruent sides.*

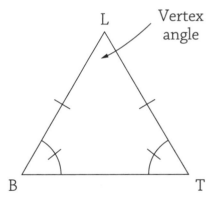

We will begin by classifying a triangle as isosceles based on the triangle's angle measurements.

EXAMPLE

▶ In the accompanying diagram of $\triangle ABC$, $\overline{AC}$ is extended to $\overline{ACD}$, $m\angle BCD = 140°$, and $m\angle A = 40°$. Show that this is an isosceles triangle.

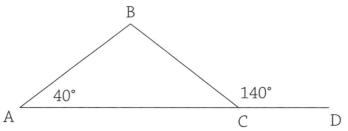

▶ $\angle BCD$ and $\angle BCA$ form a linear pair, which means their sum is 180°. That knowledge, accompanied with the fact that the sum of the three angles of a triangle equals 180°, will give us the ability to find all the measurements of the angles in the triangle and show that the triangle is isosceles.

$$m\angle BCD + m\angle BCA = 180°$$
$$\begin{array}{r} \cancel{140} + m\angle BCA = 180 \\ \underline{-\ \cancel{140} \qquad\qquad -140} \\ m\angle BCA = 40° \end{array}$$

$$m\angle A + m\angle B + m\angle BCA = 180$$
$$40 + m\angle B + 40 = 180$$
$$\begin{array}{r} \cancel{80} + m\angle B = 180 \\ \underline{-\ \cancel{80} \qquad\quad -80} \\ m\angle B = 100° \end{array}$$

▶ $\triangle ABC$ is an isosceles triangle because it has two equal angles, $\angle BCA$ and $\angle A$. It is also an obtuse triangle because it contains one obtuse angle.

The following example uses the properties of an isosceles triangle to solve a problem algebraically.

▶ In isosceles $\triangle XYZ$, $\overline{XY} \cong \overline{YZ}$ and $m\angle XYZ = 20°$. Find the measure of $\angle X$.

▶ Drawing a diagram will always help in these situations. Because this triangle is isosceles, we know that the angles opposite the congruent sides, $\angle X$ and $\angle Z$ are equal. Placing $x$ in the diagram for your unknown will help you set up an algebraic equation to solve.

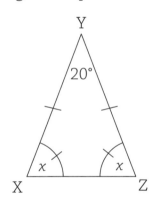

$$m\angle X + m\angle Y + m\angle Z = 180°$$
$$x + 20 + x = 180$$
$$2x + \cancel{20} = 180$$
$$\underline{-\cancel{20} \quad -20}$$
$$\frac{\cancel{2}x}{\cancel{2}} = \frac{160}{2}$$
$$x = 80$$
$$m\angle X = 80°$$

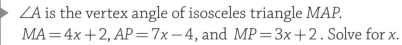

We can now look at an example that uses the property that an isosceles triangle has two equal sides.

EXAMPLE

▶ $\angle A$ is the vertex angle of isosceles triangle $MAP$.
$MA = 4x + 2$, $AP = 7x - 4$, and $MP = 3x + 2$. Solve for $x$.

▶ We are given that the triangle is isosceles, which means we know it has two equal sides. Since $A$ is the vertex, we know both equal sides must have $A$ as a common endpoint. It usually helps to draw a diagram with the given information. Notice the markings on the triangle to show the two equal sides.

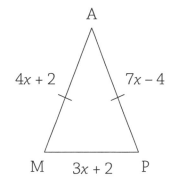

$$MA = AP$$
$$\cancel{4}x + 2 = 7x - \cancel{4}$$
$$-\cancel{4}x + 4 - 4x + \cancel{4}$$
$$\frac{6}{3} = \frac{\cancel{3}x}{\cancel{3}}$$
$$2 = x$$

**EASY MISTAKE** Don't make the common mistake of setting all three sides equal to 180°! You can see in the diagram that we need to set $MA = AP$.

An **equilateral triangle** is a triangle with all sides equal in length. We already learned that when sides of a triangle are equal the angles opposite them are equal. This means that an equilateral triangle also has all three angles equal. We can also call this triangle **equiangular**.

Our first example will incorporate the classification of an equilateral triangle by its angle measure.

**BTW**

*An equilateral triangle has three equal angles. We know that the angles of a triangle always add to 180°. This means that all three angles of an equilateral triangle must be 60°.*

**EXAMPLE**

In the accompanying diagram of $\triangle DOG$, $m\angle D = 4x$, $m\angle O = 3.5x + 7.5$, and $m\angle G = 5x - 15$. Prove algebraically that $\triangle DOG$ is an equilateral triangle.

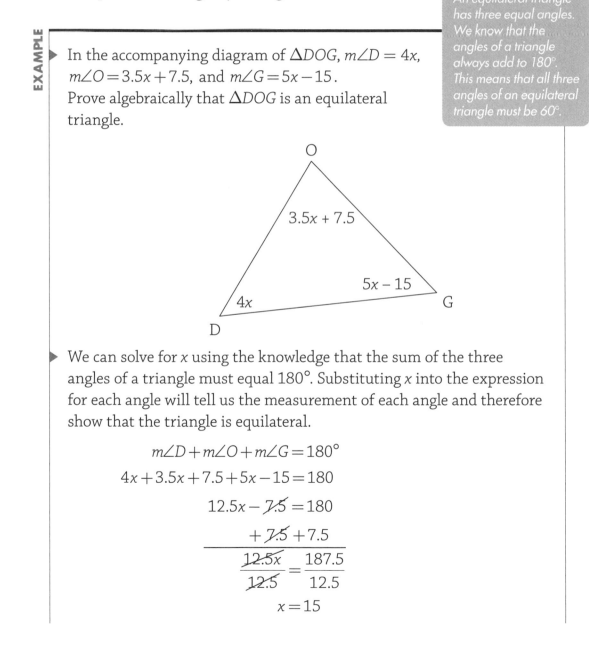

We can solve for $x$ using the knowledge that the sum of the three angles of a triangle must equal 180°. Substituting $x$ into the expression for each angle will tell us the measurement of each angle and therefore show that the triangle is equilateral.

$$m\angle D + m\angle O + m\angle G = 180°$$
$$4x + 3.5x + 7.5 + 5x - 15 = 180$$
$$12.5x - 7.5 = 180$$
$$\underline{\quad + 7.5 \quad + 7.5 \quad}$$
$$\frac{12.5x}{12.5} = \frac{187.5}{12.5}$$
$$x = 15$$

$$m\angle D = 4x = 4(15) = 60°$$
$$m\angle O = 3.5x + 7.5 = 3.5(15) + 7.5 = 60°$$
$$m\angle G = 5x - 15 = 5(15) - 15 = 60°$$

▶ $\triangle DOG$ is an equilateral triangle because it contains three equal angles.

Now let's try a problem that requires the knowledge that all sides of an equilateral triangle are equal.

**EXAMPLE**

▶ Solve for $x$ in the accompanying diagram of equilateral triangle *PIG*.

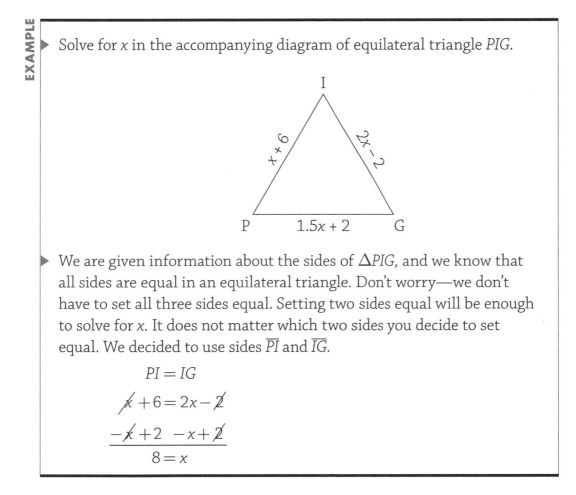

▶ We are given information about the sides of $\triangle PIG$, and we know that all sides are equal in an equilateral triangle. Don't worry—we don't have to set all three sides equal. Setting two sides equal will be enough to solve for $x$. It does not matter which two sides you decide to set equal. We decided to use sides $\overline{PI}$ and $\overline{IG}$.

$$PI = IG$$
$$\cancel{x} + 6 = 2x - \cancel{2}$$
$$\underline{-\cancel{x} + 2 \quad -x + \cancel{2}}$$
$$8 = x$$

The last triangle we will be classifying is a **scalene triangle**. A scalene triangle is a triangle with all sides unequal and all angles unequal.

The following example will show that a triangle is scalene using the measurement of the angles.

In the accompanying diagram of $\triangle NBA$, $m\angle N = 3x + 4$, $m\angle B = 4x + 5$, and $m\angle A = 9x - 5$. Show that $\triangle NBA$ is a scalene triangle.

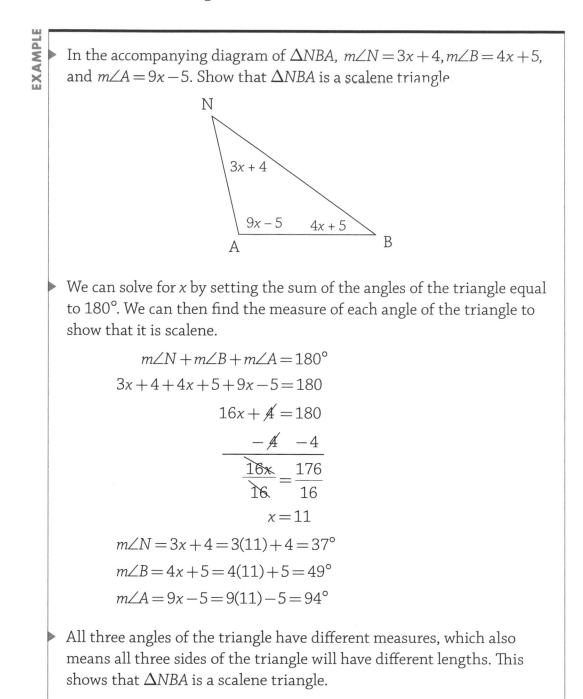

We can solve for $x$ by setting the sum of the angles of the triangle equal to $180°$. We can then find the measure of each angle of the triangle to show that it is scalene.

$$m\angle N + m\angle B + m\angle A = 180°$$
$$3x + 4 + 4x + 5 + 9x - 5 = 180$$
$$16x + \cancel{4} = 180$$
$$\underline{\phantom{16x} -\cancel{4} \quad -4}$$
$$\frac{\cancel{16x}}{\cancel{16}} = \frac{176}{16}$$
$$x = 11$$

$$m\angle N = 3x + 4 = 3(11) + 4 = 37°$$
$$m\angle B = 4x + 5 = 4(11) + 5 = 49°$$
$$m\angle A = 9x - 5 = 9(11) - 5 = 94°$$

All three angles of the triangle have different measures, which also means all three sides of the triangle will have different lengths. This shows that $\triangle NBA$ is a scalene triangle.

## Relationships of the Sides and Angles of Triangles

Let's expand on the relationship between the sides of a triangle and the angles opposite them. We already learned that if two sides of a triangle are equal, then the angles opposite them are equal. We also learned that if the angles of a triangle are all unequal, then the lengths of the sides are also unequal. This means we can arrange the sides of a triangle in size order based on their angle measurements and vice versa.

To see this firsthand, let's go back to the previous example of $\triangle NBA$ in which we found the measures of the angles to be $\angle N = 37°$, $\angle B = 49°$, and $m\angle A = 94°$. You will notice that the angles are currently in size order from smallest to largest. This means that the sides opposite those angles will also be in order from smallest to largest. The diagram might help you to see this more clearly.

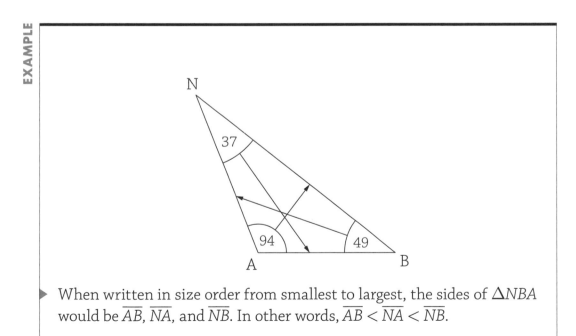

When written in size order from smallest to largest, the sides of $\triangle NBA$ would be $\overline{AB}$, $\overline{NA}$, and $\overline{NB}$. In other words, $\overline{AB} < \overline{NA} < \overline{NB}$.

The following example shows how to use the lengths of the sides of a triangle to determine the largest angle of a triangle.

**BTW**

The largest side of a triangle is always found opposite the largest angle of the triangle. This also means that the smallest angle of a triangle is found opposite the smallest side of the triangle.

**EXAMPLE**

▶ Name the largest angle in $\triangle DEF$ if $DF = 7$, $EF = 10$, and $DF = 12$.

▶ Drawing a diagram will be instructive in locating the angle opposite $\overline{DF}$, the largest side of the triangle.

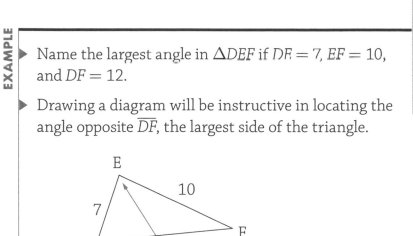

▶ $\angle E$ is the largest angle in the triangle because it is opposite the largest side of the triangle.

Did you know that you cannot always create a triangle out of three segments? The sum of the two smaller sides of a triangle must be greater than the largest side of the triangle. The following diagram shows three segments made from toothpicks. The first segment is one toothpick long, the second segment two toothpicks long, and the third segment three toothpicks long. You can see that it is impossible for all the endpoints of the toothpicks to meet to form a triangle. Right now, the largest segment measures three units while the sum of the two smaller segments also measures three units. For the endpoints to meet, we would either need to make the two smaller segments bigger or the largest segment smaller.

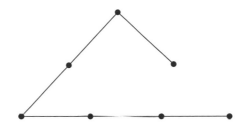

Let's look at what happens when we add a toothpick to the smallest side. Now the two smaller segments are both two toothpicks long, and the longest segment is still three toothpicks long. The endpoints of the toothpick segments can now meet to form a triangle.

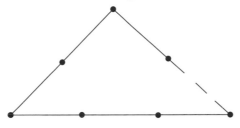

The sum of the two smaller sides is now 4, making it greater than the longest side measuring 3. This shows that segments can only form a triangle if the sum of the lengths of the two smaller segments is greater than the length of the longest segment. Using this rule, we can determine the possible lengths of a side of a triangle when given information about the other two sides. Consider the following example.

**EXAMPLE**

▶ Two sides of a triangle measure 4 and 7. What are the possible lengths of the third side of the triangle?

▶ Let's draw a diagram that contains this information. We will label the third side with an $x$.

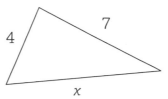

- There are two possible scenarios:
First, the two smaller sides of the triangle are 4 and 7, and the largest side is $x$. We know the sum of the two smaller sides, 4 and 7, must be larger than the largest side, $x$.

$$4 + 7 > x$$
$$11 > x$$

- Second, the two smaller sides of the triangle are 4 and $x$, and the largest side is 7. We know the sum of the two smaller sides, 4 and $x$, must be larger than the largest side, 7.

$$\begin{array}{r} \cancel{4} + x > 7 \\ -\cancel{4} \quad\;\; -4 \\ \hline x > 3 \end{array}$$

- If $x > 3$ and $x < 11$, then $3 < x < 11$.

**BTW**

*An easy way to find the possible lengths of the third side of a triangle is understanding that the third side has to be larger than the difference of the two given sides but smaller than the sum of the two given sides.*

## Median, Altitude, and Angle Bisector

A median, altitude, and angle bisector are special segments that can be drawn inside triangles that play a major role in the geometry of triangles. In the first chapter of this book, we learned that an **angle bisector** divides an angle into two congruent angles.

In a triangle, an angle bisector can be drawn from any of the three angles of a triangle to the opposite sides. In the accompanying diagram of $\Delta ABC$, you can see that angle bisector $\overline{AD}$ divides $\angle A$ into two congruent angles. We can conclude that $\angle BAD \cong \angle CAD$.

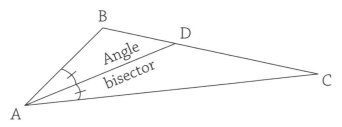

Let's look at an example in which we use the property of an angle bisector to classify a triangle.

**EXAMPLE**

► In the following diagram of $\triangle ABC$, $m\angle A = 50°$, $m\angle ABC = 100°$, $m\angle C = 30°$, and angle bisector $\overline{BD}$ is drawn. What type of triangle is triangle $ABD$? What about triangle $BDC$?

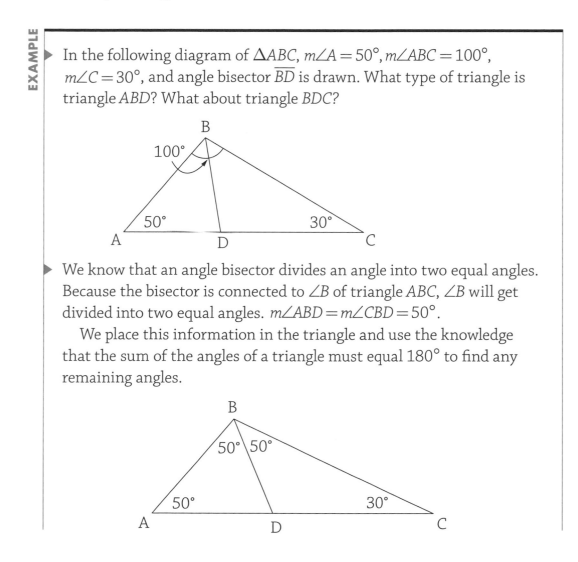

► We know that an angle bisector divides an angle into two equal angles. Because the bisector is connected to $\angle B$ of triangle $ABC$, $\angle B$ will get divided into two equal angles. $m\angle ABD = m\angle CBD = 50°$.

We place this information in the triangle and use the knowledge that the sum of the angles of a triangle must equal 180° to find any remaining angles.

$$m\angle A + m\angle ABD + m\angle BDA = 180°$$
$$50 + 50 + m\angle BDA = 180$$
$$\cancel{100} + m\angle BDA = 180$$
$$\underline{-\cancel{100} \qquad\qquad -100}$$
$$m\angle BDA = 80°$$

$$m\angle C + m\angle CBD + m\angle BDC = 180$$
$$30 + 50 + m\angle BDC = 180$$
$$\cancel{80} + m\angle BDC = 180$$
$$\underline{-\cancel{80} \qquad\qquad -80}$$
$$m\angle BDC = 100°$$

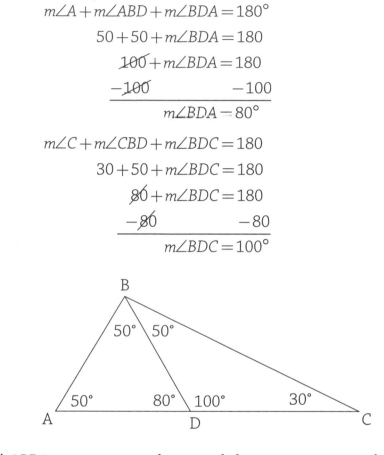

▶ $\triangle ABD$ is an acute isosceles triangle because it contains three acute angles, two of which are equal. $\triangle BDC$ is an obtuse scalene triangle because it contains one obtuse angle and three angles of different measurements.

A **median** is a segment drawn from a vertex of the triangle to the midpoint of the opposite side. Do you remember learning in the first chapter that midpoints and segment bisectors divide a segment into two congruent segments? Because a median connects to a midpoint, a median divides a segment into two congruent segments. The following diagram shows $\triangle ABC$ with a median drawn from $\angle A$. We can conclude that $\overline{BD} \cong \overline{DC}$.

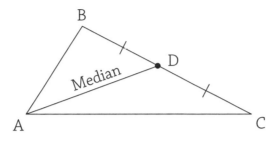

The following example shows how we could use the property of a median in a geometric proof.

**EXAMPLE**

▶ In the accompanying diagram of $\triangle TAG$, $\overline{GI}$ is a median. What can you conclude from this information?

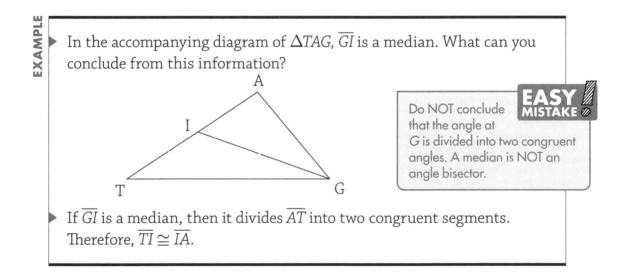

Do NOT conclude that the angle at *G* is divided into two congruent angles. A median is NOT an angle bisector.

**EASY MISTAKE**

▶ If $\overline{GI}$ is a median, then it divides $\overline{AT}$ into two congruent segments. Therefore, $\overline{TI} \cong \overline{IA}$.

Now let's look at an example in which we use the property of a median algebraically.

▶ In the accompanying diagram of $\triangle BAG$, $\overline{BE}$ is a median. The perimeter of $\triangle BAG$ is 100. If $AE = x + 6$, $AB = 3x + 1$, and $BG = 5x - 3$, what is the length of $\overline{BG}$?

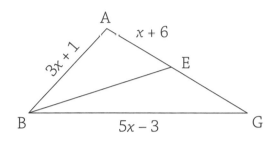

▶ $\overline{BE}$ is a median, which means it will divide $\overline{AG}$ into two equal segments, $\overline{AE}$ and $\overline{EG}$. In other words, if $AE = x + 6$, then $EG = x + 6$. We were told the perimeter is 100, which means the sum of all the sides of the triangle equals 100. That is enough information for us to set up an equation to solve for $x$.

$$3x + 1 + x + 6 + x + 6 + 5x - 3 = 100$$
$$10x + 10 = 100$$
$$\underline{\quad -10 \quad -10 \quad}$$
$$\frac{10x}{10} = \frac{90}{10}$$
$$x = 9$$

$$BG = 5x - 3$$
$$BG = 5(9) - 3$$
$$BG = 42$$

The **altitude** of a triangle is a segment that extends from any angle of a triangle and is perpendicular to a line containing the opposite side. This means that when the altitude meets the opposite side of the triangle,

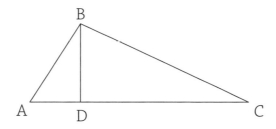

it forms right angles. *It does not mean that it is a perpendicular bisector!*
The accompanying diagram shows $\triangle ABC$ with altitude $\overline{BD}$ drawn. We can
conclude that $\angle BDA$ and $\angle BDC$ are right angles.

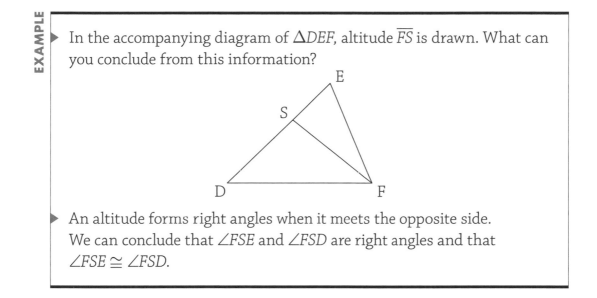

The following example shows the use of an altitude in a geometric proof.

**EXAMPLE**

▶ In the accompanying diagram of $\triangle DEF$, altitude $\overline{FS}$ is drawn. What can
you conclude from this information?

▶ An altitude forms right angles when it meets the opposite side.
We can conclude that $\angle FSE$ and $\angle FSD$ are right angles and that
$\angle FSE \cong \angle FSD$.

Now we will see an example where we use the property of an altitude to
solve an algebraic problem.

▶ In the accompanying diagram of $\triangle CAP$, altitude $\overline{AM}$ is drawn. $m\angle AMC = 4x + 22$, $m\angle C = 3x + 4$, and $m\angle CAP = 6x + 3$. Find $m\angle MAP$.

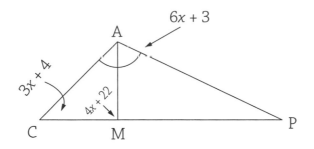

▶ If $\overline{AM}$ is an altitude, then $m\angle AMC = 90°$. This gives us the ability to solve for x, which we can plug into the expressions to find the measures of the angles given in the problem.

$$4x + 22 = 90$$
$$\underline{-22 \quad -22}$$
$$\frac{4x}{4} = \frac{68}{4}$$
$$x = 17$$
$$m\angle C = 3x + 4 = 3(17) + 4 = 55°$$
$$m\angle CAP = 6x + 3 = 6(17) + 3 = 105°$$

▶ Placing the new found information into the diagram, we can try to find the measures of other angles in the triangle.

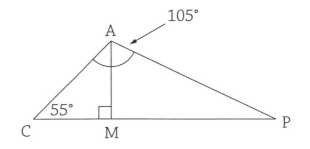

▶ Whenever our students are stuck on a triangle problem such as this one, we always tell them to look for any triangles or linear pairs because you know those angles must add to 180°. You can now see that the angles in $\triangle CAM$ must add to 180°, which will give us the ability to find $\angle CAM$. Because we already know the measure of $\angle CAP$, we will be able to subtract to find the measure of $\angle MAP$.

$$m\angle C + m\angle AMC + m\angle CAM = 180°$$
$$55 + 90 + m\angle CAM = 180$$
$$145 + m\angle CAM = 180$$
$$-\ 145 \qquad\qquad -145$$
$$\overline{\hspace{3cm}}$$
$$m\angle CAM = 35°$$

$$m\angle MAP = m\angle CAP - m\angle CAM$$
$$m\angle MAP = 105 - 35$$
$$m\angle MAP = 70°$$

> **BTW**
>
> *The median, altitude, and angle bisector of an isosceles triangle are all the same segment when coming from the vertex angle of the isosceles triangle. This means when an altitude is drawn from the vertex angle of an isosceles triangle, it is a perpendicular bisector!*

Did you read the really cool fact in the BTW?  We are going to use it to solve the following example.

**EXAMPLE**

▶ In the accompanying diagram of $\triangle ABC$, $\overline{AB} \cong \overline{BC}$, and $m\angle BAC = 68°$. If $\overline{BD}$ is the altitude of $\triangle ABC$, what is $m\angle ABD$?

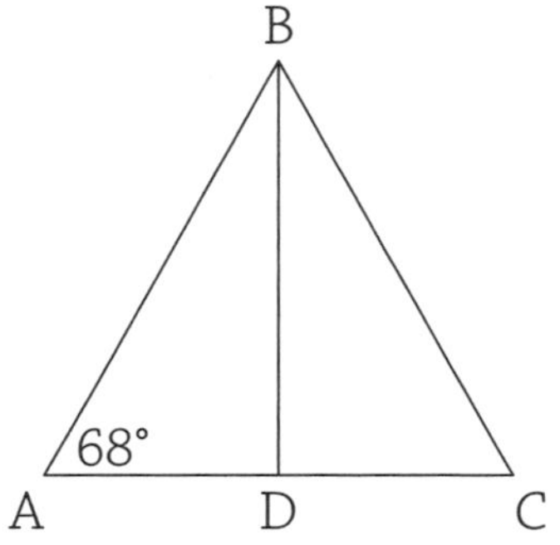

▶ We can use two concepts to answer this question. The first is the simple concept that altitudes form right angles, which means $m\angle BDA = 90°$. We can add the three angles of $\triangle ABC$ and set them equal to $180°$ to solve for $m\angle ABD$.

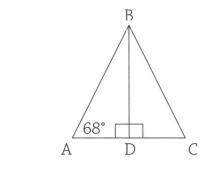

$$m\angle BDA + m\angle BAD + m\angle ABD = 180°$$
$$90 + 68 + m\angle ABD = 180$$
$$\cancel{158} + m\angle ABD = 180$$
$$\underline{\phantom{=}-158 \qquad\quad -158}$$
$$m\angle ABD = 22°$$

▶ We could also have used the concept that, when drawn from the vertex of an isosceles triangle, the altitude is also an angle bisector, which means it will divide the vertex angle into two equal angles. We know that if two sides of a triangle are congruent, then the angles opposite them are congruent. This means that $m\angle C = 68°$. We can find the measure of the vertex angle knowing that the three angles of a triangle must add to $180°$ and then divide the vertex angle in half to find $m\angle ABD$.

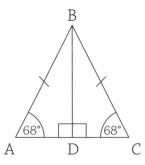

$$m\angle BAC + m\angle BCA + m\angle ABC = 180$$
$$68 + 68 + m\angle ABC = 180$$
$$\cancel{136} + m\angle ABC = 180$$
$$\underline{-\cancel{136} \qquad\qquad -136}$$
$$m\angle ABC = 44°$$

$$m\angle ABD = \frac{m\angle ABC}{2}$$
$$m\angle ABD = \frac{44}{2} = 22°$$

If the median, altitude, and angle bisector are all the same segment when drawn from the vertex angle of an isosceles triangle, then *the median, altitude, and angle bisector are all the same segment when drawn from any of the three angles of an equilateral triangle!* Let's use this concept in the following example to draw conclusions for a geometric proof.

**EXAMPLE**

▶ In the accompanying diagram, $\triangle DOE$ is an equilateral triangle with $\overline{DN}$ as the median. What conclusions can you draw about $\triangle DOE$?

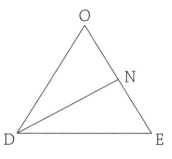

▶ The median of a triangle divides a segment into two congruent segments, which means we can conclude that $\overline{ON} \cong \overline{NE}$.

▶ In an equilateral triangle, the median is also the altitude, which means we can conclude that $\angle DNE$ and $\angle DNO$ are right angles. This also means that $\angle DNE \cong \angle DNO$ because all right angles are congruent.

▶ Our last conclusion stems from the fact that in an equilateral triangle the median is also the angle bisector. This means that it divides the angle into two congruent angles. We can therefore conclude that $\angle NDO \cong \angle NDE$.

## EXERCISES

### EXERCISE 3-1

*Answer each of the following questions.*

1. The perimeter of isosceles triangle $ABC$ with the vertex at $\angle B$ is 42. If $AB = x + 6$ and $AC = 4x - 30$, find the length of $\overline{AC}$.

2. In $\triangle SIT$, $m\angle S = 3x + 9$, $m\angle I = 8x + 2$, and $m\angle T = 5x - 7$. Classify this triangle as isosceles, equilateral, right, or scalene.

3. In $\triangle RED$, $m\angle R = 2x + 4$, $m\angle E = 3x + 5$, and $m\angle D = 13x + 9$. Classify this triangle as either acute, obtuse, or right.

4. In equilateral triangle $ABC$, $m\angle C = 7x + 4$. Solve for $x$.

5. In $\triangle CAT$, $m\angle C = 6x + 5$, $m\angle A = 15x + 8$, and $m\angle T = 7x - 1$. Classify this triangle as isosceles, equilateral, right, or scalene.

6. Name the largest angle in $\triangle ABC$ if $AB = 12$, $BC = 10$, and $AC = 8$.

7. Name the smallest angle in $\triangle MIT$ if $MI = 16$, $IT = 9$, and $MT = 20$.

8. In $\triangle RST$, $m\angle R = 35°$, $m\angle S = 70°$, and $m\angle T = 75°$. List the sides of the triangle in order from smallest to largest.

9. In $\triangle ABC$, $AB = 6$, and $BC = 10$. Find the largest possible integer value for the length of $\overline{AC}$.

10. In $\triangle MNO$, $MN = 12$ and $NO = 9$. Find the smallest possible integer value for the length of $\overline{MO}$.

**EXERCISE 3-2**

*Answer each of the following questions.*
*Use the given diagrams to help you answer these questions.*

1. Solve for $x$ in the accompanying diagram of $\triangle ABC$ where $m\angle A = x$, $m\angle B = 3x + 15$, and $m\angle C = 4x + 5$.

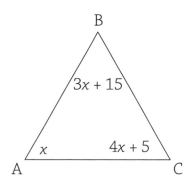

2. In the accompanying diagram of $\triangle DEF$, $m\angle D = 9x$, $m\angle E = 4x - 6$, and $m\angle F = 3x - 6$. Find $m\angle E$.

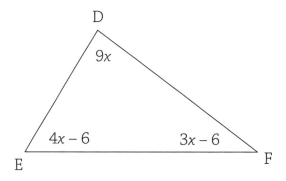

3. In the accompanying diagram of $\triangle PQR$, $\overline{PR}$ is extended to $\overline{PRS}$. If $m\angle QRS = 110°$ and $m\angle P = 80°$, what is the degree measure of $\angle Q$?

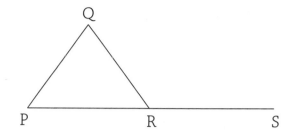

4. In the accompanying diagram of $\triangle MAD$, $\overline{MD}$ is extended to $\overline{MDE}$, $m\angle M = 4x + 9$, $m\angle A = 5x + 1$, and $m\angle ADE = 10x - 4$. Find $m\angle ADE$.

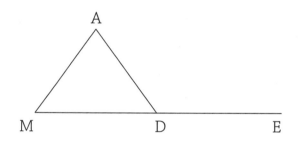

5. In the accompanying diagram of $\triangle DOG$, $m\angle D = 4x - 6$, $m\angle O = 6x - 1$, and $m\angle G = 5x + 7$. Name the largest angle in $\triangle DOG$.

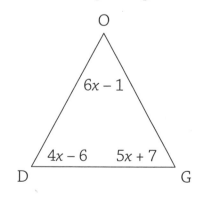

6. In the accompanying diagram of $\triangle PQR$, $\overline{QS}$ is an angle bisector. What can you conclude from this given information?

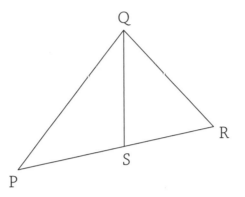

7. In the accompanying diagram of $\triangle RST$, $\overline{RL}$ is a median. What can you conclude from this given information?

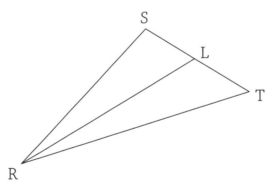

8. In the accompanying diagram of $\triangle LAS$, altitude $\overline{AT}$ is drawn. State a valid conclusion from this given information.

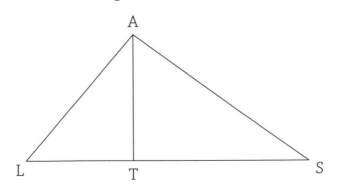

9. In the accompanying diagram of $\triangle ABC$, altitude $\overline{AD}$ is drawn. If $m\angle C = 25°$, what is $m\angle DAC$?

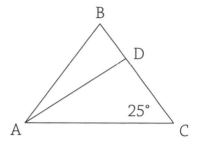

10. In the accompanying diagram of isosceles triangle $DEF$, altitude $\overline{EG}$ is drawn from vertex $\angle E$. $DE = x + 6$, $DF = 3x$, and $EG = x + 4$. If the perimeter of $\triangle DEF = 32$, find the perimeter of $\triangle DEG$.

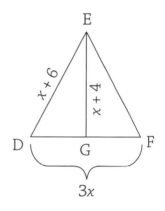

# 4 Centers of a Triangle

## MUST KNOW

⚡ Medians, angle bisectors, altitudes, and perpendicular bisectors create the different centers of a triangle: *centroid, incenter, orthocenter,* and *circumcenter.*

⚡ The centroid of a triangle will always lie inside the triangle.

⚡ The Euler line is a unique line in a triangle that includes the centroid, circumcenter, and orthocenter.

**C**an you find the exact center of a triangle? Is there a special way to find it? What would be the name of this point? In this chapter, we will investigate the four different triangle centers: **centroid, incenter, orthocenter**, and **circumcenter**. These points are found where special lines of a triangle intersect. Some of our students find it challenging to remember the names of the different triangle centers, so they created a nice little trick: **I O**wn **C**redit **C**ards! The first letter of each word is the same as the first letter in each of the center names. Now let's learn what makes each term so special.

## Centroid of a Triangle

The **centroid** of a triangle is found where the three medians of the triangle coincide. We have already learned that the median of a triangle is drawn from a vertex of the triangle to the midpoint of the opposite side of the triangle. The centroid of a triangle will always lie inside the triangle.

In the following diagram of $\Delta RST$, point $C$ represents the centroid because the three medians of the triangle intersect at that point.

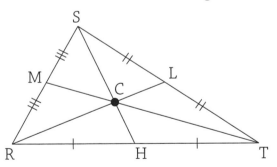

**BTW**

The centroid is often known as the **center** of gravity of a triangle.

The centroid of a triangle divides each median of the triangle into segments with a 2:1 ratio. $\overline{TC}:\overline{CM}$, $\overline{RC}:\overline{CL}$, and $\overline{SC}:\overline{CH}$ are all 2:1. This means that if $SC = 20$, then $\overline{CH}$ is half of that, which is 10.

$$\frac{\overline{SC}}{\overline{CH}} = \frac{2}{1}$$

$$\frac{20}{\overline{CH}} = \frac{2}{1}$$

$$\frac{20}{2} = \frac{\cancel{2}CH}{\cancel{2}}$$

$$10 = CH$$

What would $\overline{TC}$ equal if $CM = 8$?

$$\frac{\overline{TC}}{\overline{CM}} = \frac{2}{1}$$

$$\frac{TC}{8} = \frac{2}{1}$$

$$TC = 16$$

## EXTRA HELP

We would set this up a little different than the other two questions because, instead of knowing the length of the pieces that make up the median, we know the entire length of the median. We know that $\dfrac{\overline{RC}}{\overline{CL}} = \dfrac{2}{1}$, so if we let $CL = x$, $\overline{RC}$ must equal $2x$. We can then set up an equation to solve for $x$.

$$\overline{RC} + \overline{CL} = 12$$

$$2x + x = 12$$

$$\frac{\cancel{3}x}{\cancel{3}} = \frac{12}{3}$$

$$x = 4$$

$$RC = 2x$$

$$RC = 2(4)$$

$$RC = 8$$

Now that we understand the ratio that is created from the centroid of a triangle, let's use the diagram from the previous example to determine. What $\overline{RC}$ would equal if $RL = 12$.

## The Incenter of a Triangle

The **incenter** of a triangle is found where the three angle bisectors of a triangle coincide. We have already learned that an angle bisector divides an angle into two congruent angles. This is why you see markings of angle congruence at each of the vertex angles of the triangle.

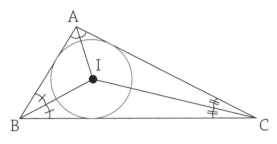

The incenter of a triangle is always located inside the triangle. It is equidistant from each side of the triangle and is the center of the circle inscribed in the triangle. This is shown in the following diagram of $\triangle ABC$.

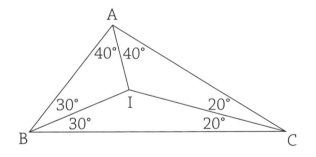

If $m\angle BAC = 80°$, then $m\angle IAB = m\angle IAC = 40°$.

If $m\angle ABC = 60°$, then $m\angle IBA = m\angle IBC = 30°$.

If $m\angle ACB = 40°$, then $m\angle ICA = m\angle ICB = 20°$.

The following example shows how to use your knowledge of an incenter to solve for an angle within the triangle.

EXAMPLE

► In the accompanying diagram of $\triangle DOG$, point $I$ represents the incenter. If $m\angle ODI = 29°$ and $m\angle IGO = 42°$, what is $m\angle OIG$?

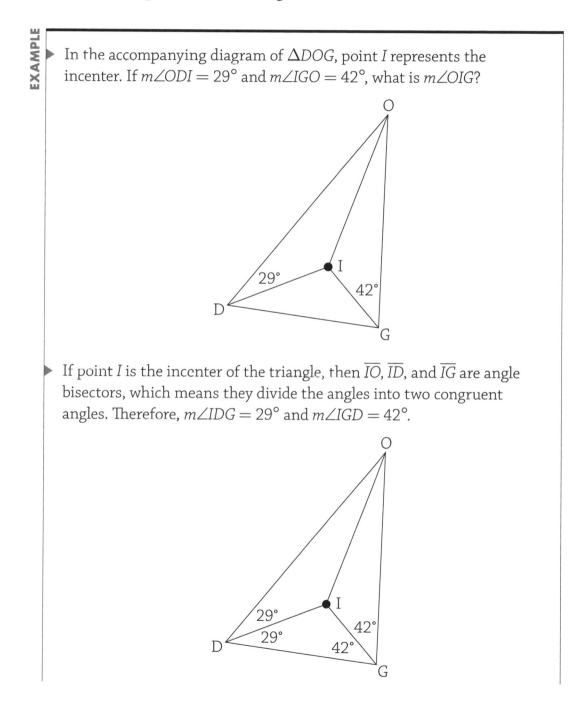

► If point $I$ is the incenter of the triangle, then $\overline{IO}$, $\overline{ID}$, and $\overline{IG}$ are angle bisectors, which means they divide the angles into two congruent angles. Therefore, $m\angle IDG = 29°$ and $m\angle IGD = 42°$.

▶ We can solve for $m\angle DOG$ using the fact that the sum of the angles of $\triangle DOG = 180°$.

$$m\angle ODG + m\angle DOG + m\angle OGD = 180°$$
$$58 + m\angle DOG + 84 = 180$$
$$\cancel{142} + m\angle DOG = 180$$
$$\underline{-\cancel{142} \qquad\qquad -142}$$
$$m\angle DOG = 38°$$

▶ We can solve for $m\angle IOG$ using our knowledge that an angle bisector divides an angle into two congruent angles.

$$m\angle IOG = \frac{m\angle DOG}{2}$$
$$m\angle IOG = \frac{38°}{2}$$
$$m\angle IOG = 19°$$

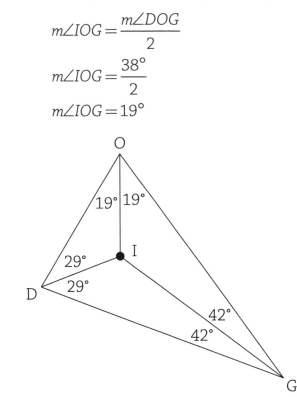

The angles formed at the incenter of a triangle ARE NOT congruent. This is only true in an equilateral triangle.

**EASY MISTAKE**

▶ We can solve for $m\angle OIG$ using the fact that the sum of the angles of $\triangle OIG = 180°$.

$$m\angle IOG + m\angle OGI + m\angle OIG = 180°$$
$$19 + 42 + m\angle OIG = 180$$
$$\cancel{61} + m\angle OIG = 180$$
$$\underline{-\cancel{61} \qquad\qquad -61}$$
$$m\angle OIG = 119°$$

# The Orthocenter of a Triangle

The **orthocenter** of a triangle is found where the three altitudes of the triangle coincide. We have already learned that the altitude of a triangle is a segment drawn from the vertex of the triangle perpendicular to the line containing the opposite side of the triangle. The orthocenter is *not* always inside the triangle. When do you think the orthocenter might lie on the outside of the triangle? Can it ever be right on the triangle? Look at the following three figures and see where the orthocenter lies for an acute triangle, a right triangle, and an obtuse triangle:

**BTW**

*The centroid and orthocenter are the same point when the triangle is equilateral.*

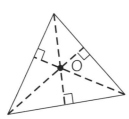

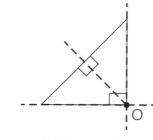

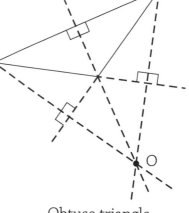

Acute triangle        Right triangle        Obtuse triangle

The following example requires the knowledge of an orthocenter to solve for a variable.

▶ In the accompanying diagram of $\triangle MAP$, point $O$ is the orthocenter. If $m\angle MNO = 2x + 20$, what is $x$?

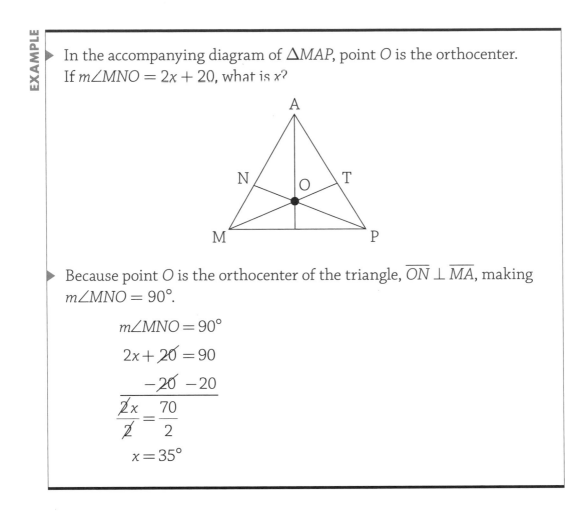

▶ Because point $O$ is the orthocenter of the triangle, $\overline{ON} \perp \overline{MA}$, making $m\angle MNO = 90°$.

$$m\angle MNO = 90°$$
$$2x + 20 = 90$$
$$\underline{-20 \quad -20}$$
$$\frac{2x}{2} = \frac{70}{2}$$
$$x = 35°$$

## The Circumcenter of a Triangle

The **circumcenter** of a triangle is found where the three perpendicular bisectors of the triangle coincide. We have already learned that a perpendicular bisector divides a segment into two congruent segments and forms right angles at the midpoint. The circumcenter is equidistant from the vertices of the triangle, which makes it the center of the circle that passes

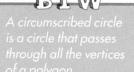

through all three vertices. The following diagram shows point *P*, the circumcenter of △*ABC*.

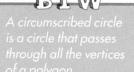

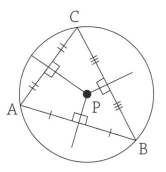

The following example requires us to use our knowledge of a circumcenter to solve for an angle.

**EXAMPLE**

▶ In the accompanying diagram of △*JOY*, point *C* represents the circumcenter. Find *m*∠*JAC*.

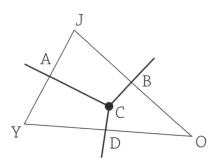

▶ If point *C* is the circumcenter of the triangle and a circumcenter is found where the perpendicular bisectors of the triangle intersect, then $\overline{CA} \perp \overline{JY}$. Therefore, *m*∠*JAC* is 90° because perpendicular lines form right angles.

Now let's use our knowledge of a circumcenter to find the lengths of segments in a triangle.

► In the accompanying diagram of $\triangle JOY$, point $C$ represents the circumcenter. If $JY = 15$, $OD = 10$, and $JB = 8$, find the measure of $\overline{AY}$ and find the perimeter of $\triangle JOY$.

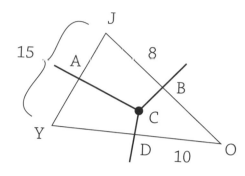

► If point $C$ is the circumcenter of the triangle and a circumcenter is found where the perpendicular bisectors of the triangle intersect, then $\overline{AY} \cong \overline{AJ}$ because a bisector divides a segment into two congruent segments. Therefore, $AY = \dfrac{JY}{2} = \dfrac{15}{2} = 7.5$.

The perimeter of a triangle is the sum of the three sides of the triangle. We already know that $JY = 15$, but we need to find the length of $\overline{OY}$ and $\overline{OJ}$. If a perpendicular bisector divides a segment into two congruent segments and $\overline{BC}$ and $\overline{CD}$ are perpendicular bisectors, then $JB = BO = 8$ and $YD = DO = 10$, as shown in the following diagram.

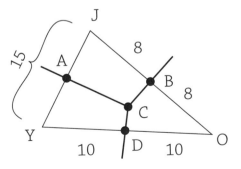

► We now know the length of all three sides of the triangle: $JY = 15$, $JO = 16$, and $OY = 20$. The perimeter of $\triangle JOY$ is therefore $15 + 16 + 20 = 51$.

Now that you have learned about the four centers of a triangle, let's see if you can recognize and name the triangle centers drawn in the diagrams below.

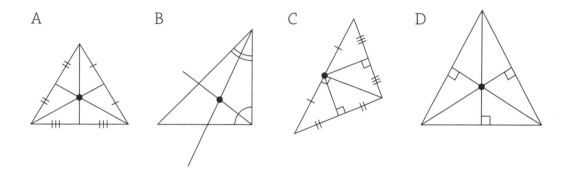

In diagram A, the markings show that the sides of the triangle have been divided into two congruent segments at the midpoint. This means that the segments drawn from the vertices of the triangle are medians. Therefore, the center of this triangle is a centroid.

In diagram B, the markings show that the angles of the triangle have been divided into two congruent angles. This means that the segments drawn from the vertices of the triangle are angle bisectors. Therefore, the center of this triangle is an incenter.

In diagram C, we see markings that show that the sides of the triangle have been divided into two congruent segments and that they have formed right angles at the midpoint. This means that the segments drawn from the vertices of the triangle are perpendicular bisectors. Therefore, the center of this triangle is a circumcenter.

In diagram D, the markings show that the segments drawn from the vertices of the triangle are perpendicular to the opposite side. This means that the segments that are drawn are altitudes. Therefore, the center of this triangle is an orthocenter.

## The Euler Line

The **Euler line** is the line that contains the centroid, circumcenter, and orthocenter of a triangle. To find the equation of the Euler line, you need to know two of these points.

**IRL** The Euler line was named after Swiss mathematician Leonhard Euler. *Euler* is pronounced "Oiler."

Let's try to find the equation of the Euler line.

**EXAMPLE**

▶ The orthocenter of a triangle is (–1,2), and the circumcenter of the triangle is (4,5). What is the equation of the Euler line?

▶ The Euler line contains the centroid, circumcenter, and orthocenter of a triangle.

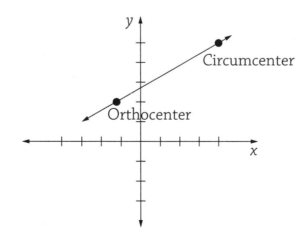

▶ To find the equation of a line, we first need to find the slope of the line that contains two of these points.

$$m = \frac{y_2 - y_1}{x_2 - x_1} = \frac{5-2}{4-(-1)} = \frac{3}{5}$$

▶ Next, we use the point-slope formula to find the equation of the Euler line:

$$m = \frac{3}{5} \text{ and } (-1, 2)$$

$$y - y_1 = m(x - x_1)$$

$$y - 2 = \frac{3}{5}(x - (-1))$$

$$y = \frac{3}{5}x + \frac{13}{5}$$

## EXERCISES

### EXERCISE 4-1

*Answer the following questions about the centers of triangles.*

1. Name the point where the medians of a triangle coincide.

2. The orthocenter is on the vertex of the triangle. What type of triangle is this?

3. True or False: The Euler line always contains the incenter, orthocenter, and circumcenter.

4. The equation of the Euler line is $y = \frac{2}{3}x + 4$. The coordinate of the orthocenter is $(6, y)$. What is the value of $y$?

5. Which center of a triangle does point $O$ represent in the following diagram?

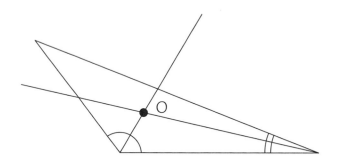

## EXERCISE 4-2

*To answer these questions, use the accompanying diagram of isosceles triangle △ISO, where ∠O ≅ ∠S, m∠CSI = 20°, and point C represents the incenter of the triangle.*

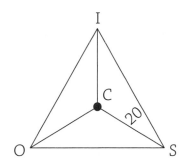

1. Find $m\angle CSO$.

2. Find $m\angle CIS$.

3. Find $m\angle SCI$.

## EXERCISE 4-3

*Use the accompanying diagram of △NIC with point P representing the incenter of the triangle to answer the following two questions.*

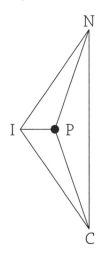

1. $m\angle PCN = 34°$, find $m\angle PCI$.

2. $m\angle PIN = 48°$ and $m\angle NIC = 2x - 10$. Solve for $x$.

## EXERCISE 4-4

*Use the accompanying diagram of △CEN with point A representing the centroid of the triangle to answer this group of questions.*

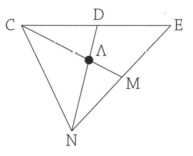

1. $MN = 8.2$. Find $ME$.
2. $NE = 22.5$. Find $MN$.
3. $CA = 17$. Find $MA$.
4. $CM = 54$. Find $CA$.
5. $DA = 14$ and $NA = 4x - 4$. Solve for $x$.
6. $DN = 81$. Find $DA$.

## EXERCISE 4-5

*Use the accompanying diagram to answer these questions.*

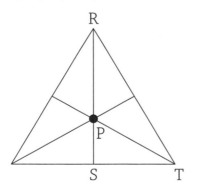

1. If point $P$ represents the orthocenter, what is $m\angle RST$?

2. If point $P$ represents the orthocenter and $m\angle RST = 3x + 30$, what is $x$?

Flashcard App

# Similarity

 **MUST KNOW**

 Similar triangles have corresponding angles that are equal in measure and corresponding sides that are in proportion.

Similar triangles have corresponding angles that are equal in measure and corresponding sides that are in proportion.

When an altitude is drawn to the hypotenuse of a right triangle, it forms three similar right triangles.

The Angle-Angle-Angle theorem, also called the Angle-Angle theorem, allows us to prove triangles are similar.

e've already learned about the congruence of triangles; it is time to learn about the similarity of triangles. If you think about it, we use the word *similar* all the time in everyday life. "The two Italian restaurants have a *similar* menu," or after hearing about a friend's experience at a new salon, you might respond with, "I had a *similar* experience." Does this mean that the two Italian restaurants have the same exact menu? Or that you had the same exact experience at the salon as your friend? No. It means that they are alike but not identical. The same is true in geometry. When we talk about two polygons being **similar**, we are saying that the polygons have maintained the same shape but are not equal in size.

This chapter will focus on the similarity of triangles. This means that we will be looking at triangles that have maintained their shape but not the lengths of their sides.

This occurs when we dilate a triangle. Dilations preserve angle measure but not distance. This means that when we dilate a triangle we create a triangle similar to the original triangle. To get an understanding of this, refer to the following diagram. You can see that both triangles are equilateral triangles. They both have three 60° angles, but one triangle has three equal sides measuring three units long and the other has three equal sides measuring six units long. They are the same shape but not equal in size. This means that they are similar.

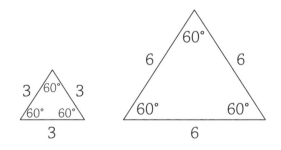

**IRL** Architects use scale models when they are taking on a project for a large building or skyscraper. These scale models are the same shape as the true building but are smaller in size. This means they are similar.

# Proportions in Similar Triangles

When two triangles are similar, their corresponding angles are congruent and their corresponding sides are in proportion. To determine which angles correspond to one another, use the order of the vertices in the names of the triangles. The symbol for similarity is ~, so if we say $\triangle ABC \sim \triangle DEF$ this means $\angle A \cong \angle D$, $\angle B \cong \angle E$, and $\angle C \cong \angle F$. It also means that $\overline{AB}$ is proportional to $\overline{DE}$, $\overline{BC}$ is proportional to $\overline{EF}$, and $\overline{AC}$ is proportional to $\overline{DF}$. Let's use the following diagram of $\triangle ABC \sim \triangle DEF$ to discuss the proportion of the sides more in depth.

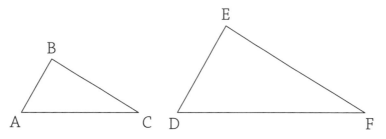

There are many different proportions that can be set up to solve for the lengths of sides in similar triangles. Be consistent! We often have our students say things to themselves to help them set up their proportions. You might hear "Left side of the little triangle over bottom of the little triangle equals left side of the big triangle over bottom of the big triangle." For the triangles in our diagram, this would look like:

$$\frac{AB}{AC} = \frac{DE}{DF}$$

Perhaps you decided to do "Left side of the little triangle over left side of the big triangle equals right side of little triangle over right side of the big triangle." For the triangles in our diagram, this would look like:

$$\frac{AB}{DE} = \frac{BC}{EF}$$

There are many more proportions that could be set up. Just remember to keep all corresponding parts in order within the proportion.

The following example shows how to use a proportion to solve for the length of a side of a triangle.

▶ The accompanying diagram shows $\triangle ABC \sim \triangle DEF$. If $AB = 3$, $DE = 12$, and $DF = 20$, what is the length of $\overline{AC}$?

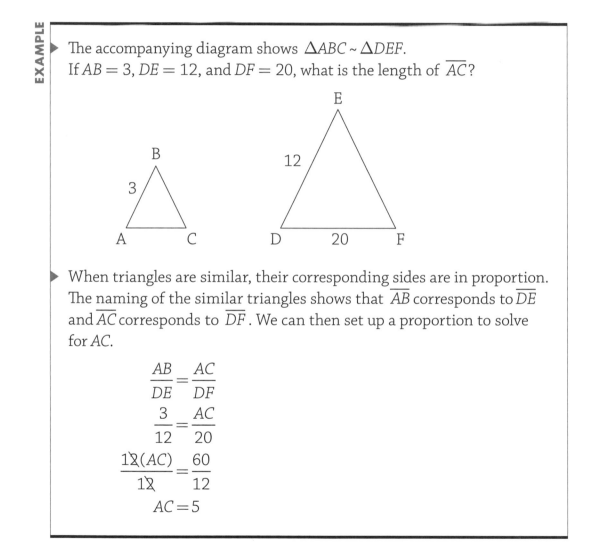

▶ When triangles are similar, their corresponding sides are in proportion. The naming of the similar triangles shows that $\overline{AB}$ corresponds to $\overline{DE}$ and $\overline{AC}$ corresponds to $\overline{DF}$. We can then set up a proportion to solve for $AC$.

$$\frac{AB}{DE} = \frac{AC}{DF}$$

$$\frac{3}{12} = \frac{AC}{20}$$

$$\frac{\cancel{12}(AC)}{\cancel{12}} = \frac{60}{12}$$

$$AC = 5$$

The following example shows why it is so important to read the order of the vertices in the names of the triangles to determine the sides that are in proportion.

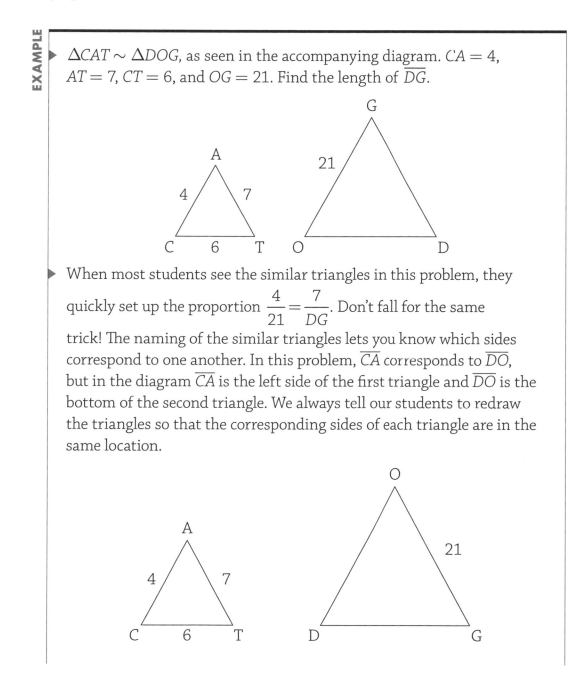

▶ $\triangle CAT \sim \triangle DOG$, as seen in the accompanying diagram. $CA = 4$, $AT = 7$, $CT = 6$, and $OG = 21$. Find the length of $\overline{DG}$.

▶ When most students see the similar triangles in this problem, they quickly set up the proportion $\dfrac{4}{21} = \dfrac{7}{DG}$. Don't fall for the same trick! The naming of the similar triangles lets you know which sides correspond to one another. In this problem, $\overline{CA}$ corresponds to $\overline{DO}$, but in the diagram $\overline{CA}$ is the left side of the first triangle and $\overline{DO}$ is the bottom of the second triangle. We always tell our students to redraw the triangles so that the corresponding sides of each triangle are in the same location.

▶ Now we can set up a proportion to solve for $DG$:

$$\frac{AT}{CT} = \frac{OG}{DG}$$

$$\frac{7}{6} = \frac{21}{DG}$$

$$\frac{\cancel{7}(DG)}{\cancel{7}} = \frac{126}{7}$$

$$DG = 18$$

## Determining Whether Triangles Are Similar

The most commonly used method for proving similarity of triangles is the **Angle-Angle-Angle (AAA) Theorem**. This theorem states that if three angles of one triangle are congruent to three corresponding angles in another triangle, the triangles are similar. You can see this in the following diagram.

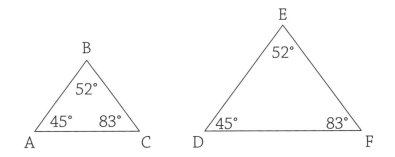

You can see that $\triangle ABC \sim \triangle DEF$ because the three corresponding angles are congruent.

The second method for proving similarity in triangles is the **Side-Side-Side (SSS) Similarity Theorem**. This method states that if the three pairs of corresponding sides of the triangles are *in proportion*, then the triangles are similar. You can see this in the following diagram.

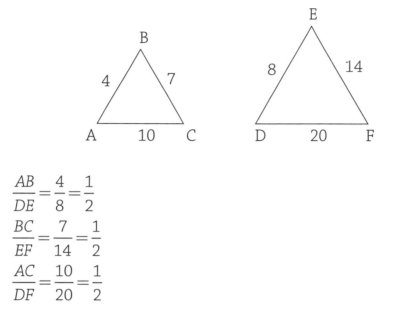

$$\frac{AB}{DE} = \frac{4}{8} = \frac{1}{2}$$

$$\frac{BC}{EF} = \frac{7}{14} = \frac{1}{2}$$

$$\frac{AC}{DF} = \frac{10}{20} = \frac{1}{2}$$

You can see that $\triangle ABC \sim \triangle DEF$ because three pairs of corresponding sides are in proportion.

The third method for proving triangle similarity is the **Side-Angle-Side (SAS) Similarity Theorem**. This theorem states that if two pairs of corresponding sides are in proportion and the included angles between these sides are congruent, then the triangles are similar. You can see this in the following diagram.

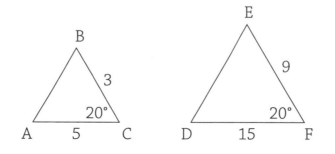

$$\frac{BC}{EF} = \frac{3}{9} = \frac{1}{3}$$

$$\frac{AC}{DF} = \frac{5}{15} = \frac{1}{3}$$

$$\angle C \cong \angle F$$

You can see that $\triangle ABC \sim \triangle DEF$ because two pairs of corresponding sides are in proportion and the angles included between these sides are congruent.

## Perimeter and Area of Similar Triangles

It is now time to explore some other properties of similar triangles. The following diagram shows triangles $RST$ and $XYZ$.

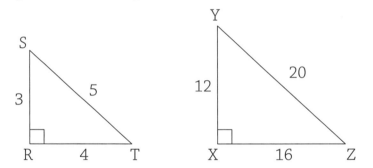

Are the triangles similar? *Yes!* We know that if the triangles are similar, then the ratio of the corresponding sides must all be the same. In other

words, $\dfrac{RS}{XY} = \dfrac{ST}{YZ} = \dfrac{RT}{XZ}$.

$$\frac{3}{12} = \frac{5}{20} = \frac{4}{16}$$

$$\frac{1}{4} = \frac{1}{4} = \frac{1}{4}$$

What is the ratio of the perimeters of $\triangle RST$ and $\triangle XYZ$?

$$\frac{\text{Perimeter of } \triangle RST}{\text{Perimeter of } \triangle XYZ} = \frac{12}{48} = \frac{1}{4}$$

You just discovered a critically important property of similar triangles: If two triangles are similar, then the ratio of the perimeters of the similar triangles is equal to the ratio of the corresponding sides of the similar triangles.

Do you think the ratio of the areas of two similar triangles will be equal to the ratio of the corresponding sides of the similar triangles? Let's go back to our diagram of triangles $RST$ and $XYZ$ to find out. Using the formula for the area of a triangle, $A = \dfrac{1}{2}$(base)(height), determine the ratio of the areas of $\triangle RST$ and $\triangle XYZ$.

$$\frac{\text{Area of } \triangle RST}{\text{Area of } \triangle XYZ} = \frac{\dfrac{1}{2}(4)(3)}{\dfrac{1}{2}(16)(12)} = \frac{6}{96} = \frac{1}{16}$$

Notice that the ratio of the areas of the triangles is *not equal* to the ratio of the corresponding sides but that it is *equal* to the square of the ratio of the corresponding sides!

Let's apply what we learned about the ratios of the perimeter and area of similar triangles to the following example.

**EXAMPLE**

▶ The accompanying diagram shows $\triangle BUG \sim \triangle CAN$, $BU = 6$ cm, and $CA = 9$ cm. The perimeter of $\triangle BUG = 20$ cm, and the area of $\triangle CAN = 36$ cm$^2$. Find the perimeter of $\triangle CAN$ and the area of $\triangle BUG$.

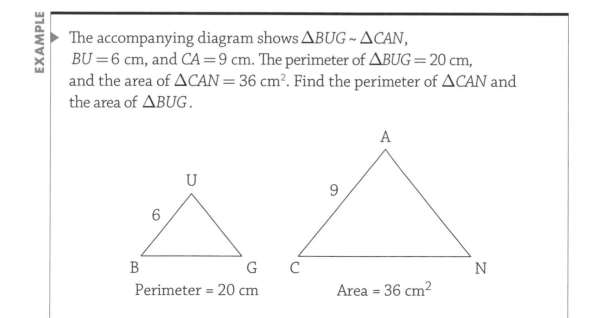

▶ The ratio of the perimeters of similar triangles is equal to the ratio of the corresponding sides.

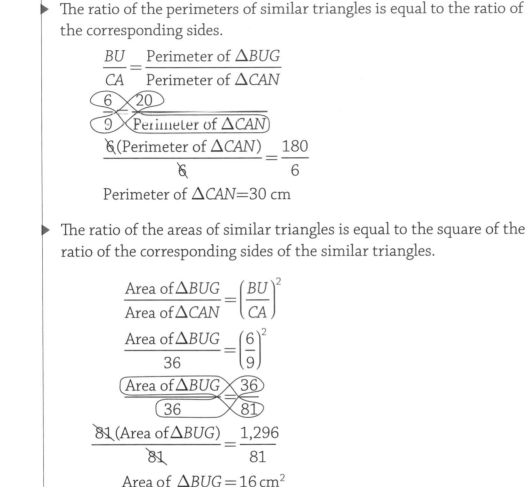

$$\frac{BU}{CA} = \frac{\text{Perimeter of } \triangle BUG}{\text{Perimeter of } \triangle CAN}$$

$$\frac{6}{9} = \frac{20}{\text{Perimeter of } \triangle CAN}$$

$$\frac{6(\text{Perimeter of } \triangle CAN)}{6} = \frac{180}{6}$$

Perimeter of $\triangle CAN = 30$ cm

▶ The ratio of the areas of similar triangles is equal to the square of the ratio of the corresponding sides of the similar triangles.

$$\frac{\text{Area of } \triangle BUG}{\text{Area of } \triangle CAN} = \left(\frac{BU}{CA}\right)^2$$

$$\frac{\text{Area of } \triangle BUG}{36} = \left(\frac{6}{9}\right)^2$$

$$\frac{\text{Area of } \triangle BUG}{36} = \frac{36}{81}$$

$$\frac{81(\text{Area of } \triangle BUG)}{81} = \frac{1{,}296}{81}$$

Area of $\triangle BUG = 16$ cm$^2$

## Parallel Lines Inside a Triangle

When a line is drawn inside a triangle and is parallel to a side of the triangle, it creates similar triangles. The parallel lines create congruent corresponding angles, which results in the three angles of the smaller triangle congruent to the three angles of the larger triangle. The triangles are similar by the AAA Similarity Theorem, which is shown in the accompanying diagram.

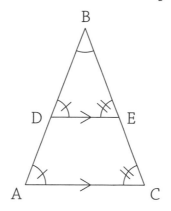

The following example uses the concept of creating similar triangles by drawing a line segment parallel to a side of a given triangle.

**EXAMPLE**

▶ In $\triangle ABC$, $D$ is a point on $\overline{AB}$ and $E$ is a point on $\overline{BC}$ such that $\overline{DE} \parallel \overline{AC}$. $BD = x + 2$, $BA = 3x - 1$, $DE = 2x + 4$, and $AC = 3x + 7$. Find the length of $AC$.

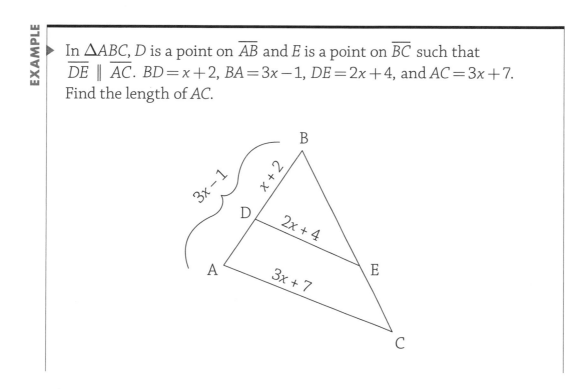

$\overline{DE}$ is parallel to $\overline{AC}$, which tells us $\triangle BDE \sim \triangle BAC$. This means that the corresponding sides of these triangles are in proportion, which will give us the ability to solve for $AC$.

$$\frac{BD}{DE} = \frac{BA}{AC}$$

$$\frac{x+2}{2x+4} = \frac{3x-1}{3x+7}$$

$$(x+2)(3x+7) = (2x+4)(3x-1)$$

$$3x^2 + 13x + 14 = 6x^2 + 10x - 4$$

$$\underline{-3x^2 - 13x - 14 \quad -3x^2 - 13x - 14}$$

$$0 = 3x^2 - 3x - 18$$

$$0 = 3(x^2 - x - 6)$$

$$0 = 3(x-3)(x+2)$$

$$x - 3 = 0 \qquad\qquad x + 2 = 0$$

$$\underline{+3 +3} \qquad\qquad \underline{-2 -2}$$

$$x = 3 \qquad\qquad x = \cancel{-2}$$

$$\text{reject}$$

$$AC = 3x + 7$$

$$AC = 3(3) + 7$$

$$AC = 16$$

The next example will require the use of the **Side Splitter Theorem**. This theorem says that if a line or line segment intersecting two sides of the triangle is drawn parallel to the third side of the triangle, then not only are the two triangles similar but the line will divide the segments of the two sides of the triangle proportionally.

EXAMPLE

▶ In the following diagram of $\triangle ABC$, $\overline{DE}$ is drawn parallel to $\overline{BC}$. $AE = 4$, $EC = 3$, and $AD = 10$. Find the length of $DB$.

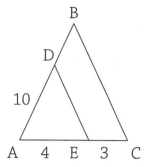

▶ $\overline{DE}$ is parallel to $\overline{BC}$, which means $\triangle ADE \sim \triangle ABC$. $\overline{DE}$ intersects sides $\overline{AB}$ and $\overline{AC}$ of triangle $ABC$. The Side Splitter Theorem says it divides those sides into proportional segments.

$$\frac{AD}{AE} = \frac{DB}{EC}$$

$$\frac{10}{4} = \frac{DB}{3}$$

$$\frac{4(DB)}{4} = \frac{30}{4}$$

$$DB = 7.5$$

▶ Technically, we did not need to use the Side Splitter Theorem to solve this problem. We could have set up a proportion with the corresponding sides of $\triangle ADE$ and $\triangle ABC$.

$$\frac{AD}{AB} = \frac{AE}{AC}$$

$$\frac{10}{10 + DB} = \frac{4}{7}$$

$$70 + 4(DB) = 70$$
$$-70 \qquad\qquad -40$$
$$\frac{4DB}{4} = \frac{30}{4}$$

$$DB = 7.5$$

A **midsegment** of a triangle is created when the midpoints of two sides of a triangle are connected. The midsegment of a triangle is parallel to the third side of the triangle and is half the length of the third side. If the third side of the triangle equals 10, then the midsegment equals 5. If the midsegment equals 12, then the third side of the triangle equals 24. The following example shows how to use a midsegment to solve for the length of a side of a triangle algebraically.

▶ In the accompanying diagram of $\triangle ABC$, $D$ is the midpoint of $\overline{AB}$ and $E$ is the midpoint of $\overline{BC}$. If $DE = \dfrac{1}{2}x + 7$ and $AC = 2x - 4$, what is the length of $\overline{AC}$?

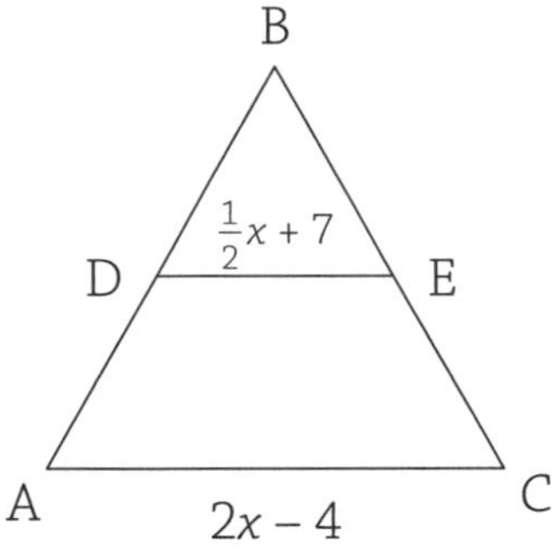

▶ The endpoints of $\overline{DE}$ are the midpoints of two sides of triangle $ABC$; therefore, $\overline{DE}$ is a midsegment. Midsegments of a triangle are equal to half the length of the third side of the triangle (the side it is parallel to).

$$DE = \frac{AC}{2}$$

$$\frac{\frac{1}{2}x + 7}{1} = \frac{2x - 4}{2}$$

$$2\left(\frac{1}{2}x + 7\right) = 2x - 4$$

$$x + 14 = 2x - \cancel{4}$$

$$\underline{-x + 4 \quad -x + \cancel{4}}$$

$$18 = x$$

$$AC = 2(18) - 4$$

$$AC = 32$$

# Proportions of Similar Right Triangles

Before we apply our knowledge of similar triangles toward geometric proofs, it is important to look at the proportions formed from similar right triangles. When an altitude is drawn in a right triangle from the right angle to the hypotenuse, it forms three similar right triangles, as shown in the accompanying diagram of right triangle *ABC*.

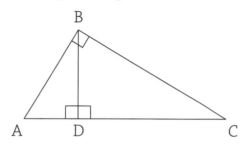

$$\triangle ABC \sim \triangle ADB \sim \triangle BDC$$

We can set up proportions with the corresponding sides of these right triangles. One proportion involves the altitude of the triangle.

$$\frac{AD}{BD} = \frac{BD}{DC}$$

$$\frac{\text{Segment of hypotenuse}}{\text{Altitude}} = \frac{\text{Altitude}}{\text{Other segment of hypotenuse}}$$

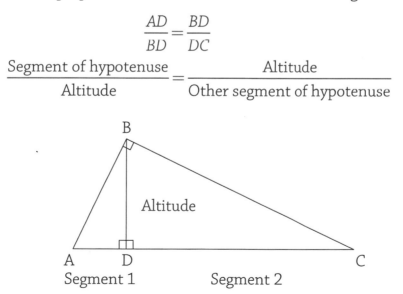

The following example uses the proportions of similar right triangles to find the length of the altitude of the triangle.

In the accompanying diagram of right triangle *RED*, altitude $\overline{ET}$ is drawn to hypotenuse $\overline{RD}$. If $RT = 3$ and $TD = 12$, what is the length of $\overline{ET}$?

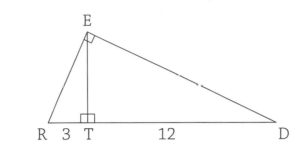

The following right triangle proportion will help us solve for the length of altitude $\overline{ET}$.

$$\frac{\text{Segment}}{\text{Altitude}} = \frac{\text{Altitude}}{\text{Other segment}}$$

$$\frac{RT}{ET} = \frac{ET}{TD}$$

$$\frac{3}{ET} = \frac{ET}{12}$$

$$\sqrt{(ET)^2} = \sqrt{36}$$

$$ET = 6$$

## EXTRA HELP

The length of the altitude is not always going to be a perfect square. If $RT = 4$ and $TD = 12$, then $ET$ would be $\sqrt{48}$. Remember from algebra, that this would simplify to $\sqrt[4]{3}$.

Another proportion that can be generated from similar right triangles involves a leg of the big right triangle, as shown in the following diagram. In this proportion, you will see the word *projection*. This is the section or segment of the hypotenuse that is closest to the leg of the right triangle.

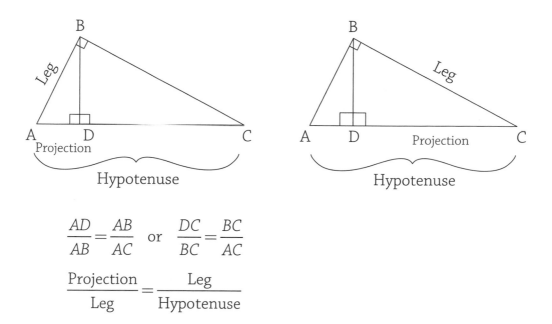

$$\frac{AD}{AB} = \frac{AB}{AC} \quad \text{or} \quad \frac{DC}{BC} = \frac{BC}{AC}$$

$$\frac{\text{Projection}}{\text{Leg}} = \frac{\text{Leg}}{\text{Hypotenuse}}$$

Let's put our understanding of this proportion to the test in the following example.

**EXAMPLE**

▶ In the accompanying diagram of right triangle $JOB$, altitude $\overline{JK}$ is drawn to hypotenuse $\overline{OB}$. If $BK = 2$ and $OK = 6$, what is the length of $\overline{JB}$?

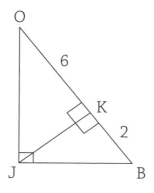

▶ Many students would instantly be confused by this diagram because the right triangle is not drawn the way they are used to seeing it. You might want to redraw the triangle by rotating it so that the altitude is vertical, as shown in the following diagram.

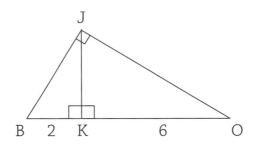

▶ To solve for $\overline{JB}$, a leg of the right triangle, we need to set up a proportion with the similar right triangles that use that leg.

$$\frac{\text{Projected segment}}{\text{Leg}} = \frac{\text{Leg}}{\text{Hypotenuse}}$$

$$\frac{BK}{JB} = \frac{JB}{BO}$$

$$\frac{2}{JB} = \frac{JB}{8}$$

$$\sqrt{(JB)^2} = \sqrt{16}$$

$$JB = 4$$

## Similar Triangle Proofs

It is time for us to apply our knowledge of similar triangles toward formal geometric proofs. We should begin by reminding ourselves what we learned at the beginning of the chapter: When two triangles are similar, all three angles of one triangle must be

**BTW**

If two triangles have two angles congruent to one another, then the third angle must also be congruent because the three angles of a triangle must always add to 180°. Therefore, when we are proving triangles similar, we normally write AA ≅ AA as our reason.

congruent to the corresponding angles of the other triangle. Thus, to prove triangles similar in a geometric proof, we will use the AAA Theorem.

Let's look at an example of a basic similar triangle proof.

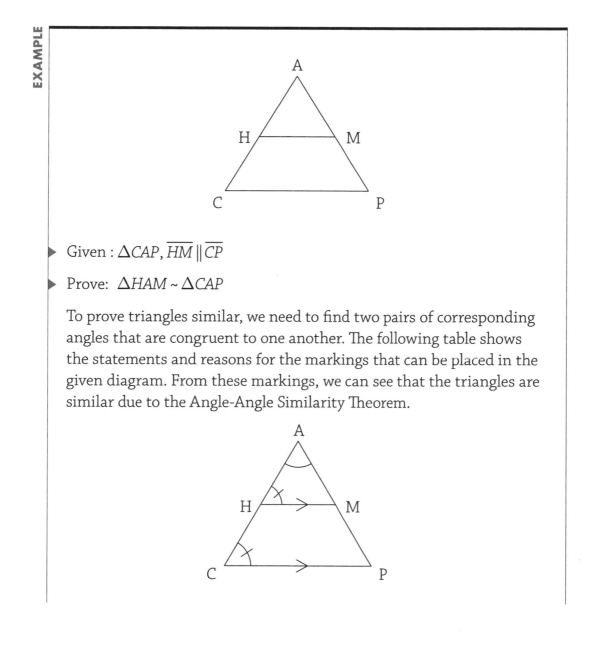

▶ Given : $\triangle CAP, \overline{HM} \parallel \overline{CP}$

▶ Prove:  $\triangle HAM \sim \triangle CAP$

To prove triangles similar, we need to find two pairs of corresponding angles that are congruent to one another. The following table shows the statements and reasons for the markings that can be placed in the given diagram. From these markings, we can see that the triangles are similar due to the Angle-Angle Similarity Theorem.

| Statement | Reason |
|-----------|--------|
| 1. $\triangle CAP$, $\overline{HM} \parallel \overline{CP}$ | 1. Given |
| 2. $\angle A \cong \angle A$ | 2. Reflexive property |
| 3. $\angle AHM \cong \angle ACP$ | 3. When parallel lines are cut by a transversal, they form congruent corresponding angles. (This concept is discussed in more detail in Chapter 7.) |
| 4. $\triangle HAM \sim \triangle CAP$ | 4. AA $\cong$ AA |

We already learned that when two triangles are similar, their corresponding sides are in proportion. Therefore, if a geometric proof asks us to prove a proportion, you first must prove that the two triangles are similar. This is shown in the following example.

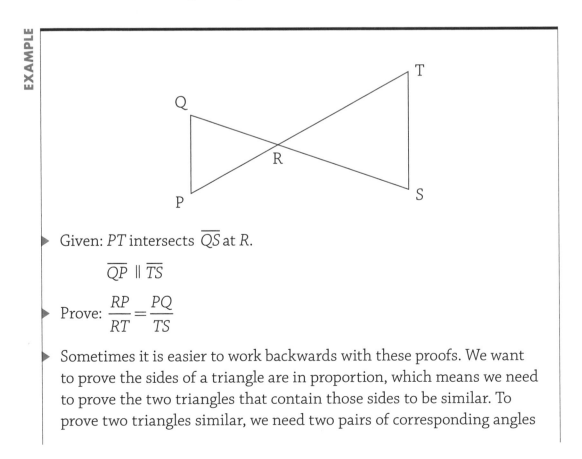

Given: $PT$ intersects $\overline{QS}$ at $R$.

$\overline{QP} \parallel \overline{TS}$

Prove: $\dfrac{RP}{RT} = \dfrac{PQ}{TS}$

Sometimes it is easier to work backwards with these proofs. We want to prove the sides of a triangle are in proportion, which means we need to prove the two triangles that contain those sides to be similar. To prove two triangles similar, we need two pairs of corresponding angles in those triangles to be congruent. Now we know to first use the given information to find two congruent angles, then state that the triangles are similar, and lastly state that the sides of the triangle are in proportion.

▶ The following table shows the statements and reasons for the markings that can be placed in the following diagram. From these markings, we can see that the triangles are similar due to the Angle-Angle Similarity Theorem and when two triangles are similar their corresponding sides are in proportion.

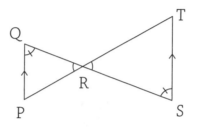

| Statement | Reason |
|---|---|
| 1. $\overline{PT}$ intersects $\overline{QS}$ at $R$. | 1. Given |
| 2. $\angle QRP \cong \angle SRT$ | 2. Vertical angles are congruent. |
| 3. $\overline{QP} \parallel \overline{TS}$ | 3. Given |
| 4. $\angle PQS \cong \angle TSQ$ | 4. When parallel lines are cut by a transversal, they form congruent alternate interior angles. |
| 5. $\triangle PQR \sim \triangle TSR$ | 5. AA $\cong$ AA |
| 6. $\dfrac{RP}{RT} = \dfrac{PQ}{TS}$ | 6. Corresponding sides of similar triangles are in proportion. |

There is one more thing we can do to build on these geometric proofs. What if we are asked to prove that the product of the segments of triangles are equal? Let's try this in the following proof.

EXAMPLE

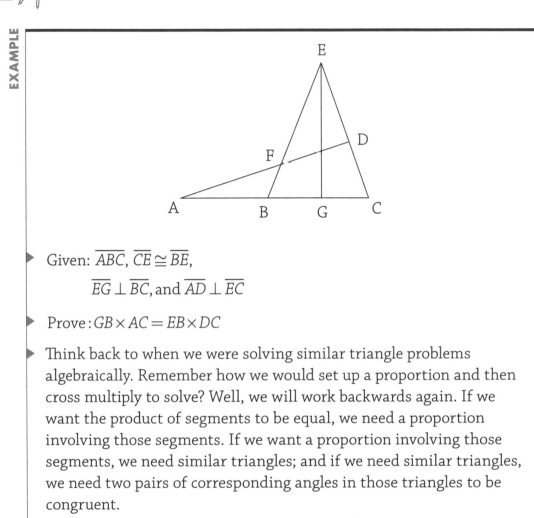

Given: $\overline{ABC}$, $\overline{CE} \cong \overline{BE}$,

$\overline{EG} \perp \overline{BC}$, and $\overline{AD} \perp \overline{EC}$

Prove: $GB \times AC = EB \times DC$

Think back to when we were solving similar triangle problems algebraically. Remember how we would set up a proportion and then cross multiply to solve? Well, we will work backwards again. If we want the product of segments to be equal, we need a proportion involving those segments. If we want a proportion involving those segments, we need similar triangles; and if we need similar triangles, we need two pairs of corresponding angles in those triangles to be congruent.

The following table shows the statements and reasons for the markings that can be placed in the following diagram. From these markings, we can see that the triangles are similar because of the Angle-Angle Similarity Theorem; when two triangles are similar, their corresponding sides are in proportion—and in a proportion, *the product of the means equals the product of the extremes.*

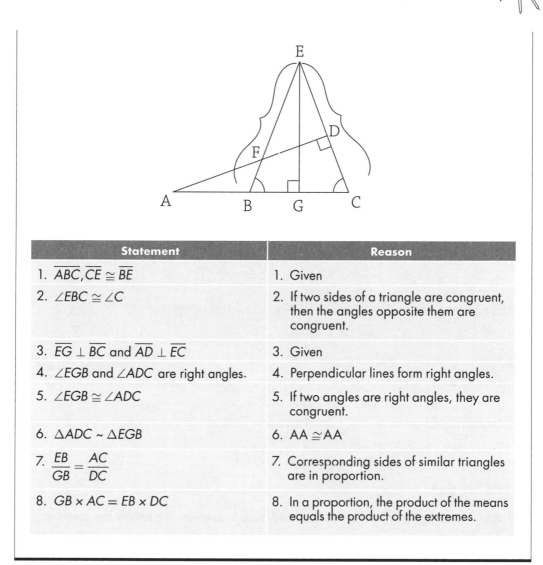

| Statement | Reason |
|---|---|
| 1. $\overline{ABC}, \overline{CE} \cong \overline{BE}$ | 1. Given |
| 2. $\angle EBC \cong \angle C$ | 2. If two sides of a triangle are congruent, then the angles opposite them are congruent. |
| 3. $\overline{EG} \perp \overline{BC}$ and $\overline{AD} \perp \overline{EC}$ | 3. Given |
| 4. $\angle EGB$ and $\angle ADC$ are right angles. | 4. Perpendicular lines form right angles. |
| 5. $\angle EGB \cong \angle ADC$ | 5. If two angles are right angles, they are congruent. |
| 6. $\triangle ADC \sim \triangle EGB$ | 6. $AA \cong AA$ |
| 7. $\dfrac{EB}{GB} = \dfrac{AC}{DC}$ | 7. Corresponding sides of similar triangles are in proportion. |
| 8. $GB \times AC = EB \times DC$ | 8. In a proportion, the product of the means equals the product of the extremes. |

## BTW

*If you are asked to prove that the product of segments are equal, you already know what the last three statements and reasons of the proof will be:*

| | |
|---|---|
| $\triangle \sim \triangle$ | $AA \cong AA$ |
| — = — | The corresponding sides of similar triangles are in proportion. |
| $(\ )(\ ) = (\ )(\ )$ | In a proportion, the product of the means equals the product of the extremes. |

## EXERCISES

### EXERCISE 5-1

*Answer the following questions.*

1. $\triangle DEF$ is the image of $\triangle ABC$ after it is dilated by a scale factor of 2 with the center of dilation at point $A$. If $m\angle A = 50°$ and $m\angle B = 67°$, what is $m\angle F$?

2. $\triangle ABC \sim \triangle DEF$. If $AC = 4$ units, $DF = 6$ units, and the area of $\triangle ABC = 9$ units$^2$, what is the area of $\triangle DEF$?

3. In right triangle $ABC$, altitude $\overline{BD}$ is drawn to hypotenuse $\overline{AC}$. If $AD = 5$ and $DC = 15$, what is the length of $\overline{AB}$?

4. $\triangle ABC \sim \triangle DEF$. $AB = 4$, $BC = 6$, and $AC = 7$. If the perimeter of $\triangle DEF$ is 51, what is the measure of the longest side of $\triangle DEF$?

5. In $\triangle RST$, $RS = 6$, $ST = 9$, and $m\angle S = 30°$. In $\triangle WXY$, $WX = 2$, $XY = 3$, and $m\angle Y = 30°$. Is $\triangle RST \sim \triangle WXY$?

### EXERCISE 5-2

*Use the given diagrams and related information to answer the following questions.*

1. $\triangle ABC \sim \triangle DEF$. $AB = 4.2$, $AC = 6.4$, and $DF = 19.2$. Find the length of $\overline{DE}$.

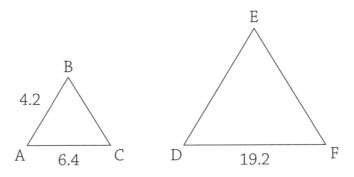

2. $m\angle CEF = 105°, m\angle A = 75°, BC = 4.2, DC = 12,$ and $DE = 10$. Find the length of $\overline{AB}$.

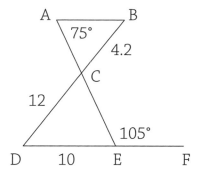

3. $\triangle DOG \sim \triangle CAT$. Find the length of $\overline{CA}$ if the perimeter of $\triangle DOG$ is 30, the perimeter of $\triangle CAT$ is 45, and $DO = 10$.

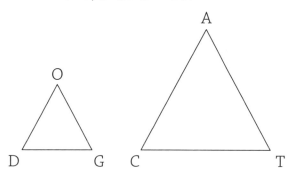

4. In $\triangle LNO$, $M$ is a point on $\overline{LN}$ and $P$ is a point on $\overline{LO}$ such that $\overline{MP} \| \overline{NO}$. $LP = x - 4$, $PO = x - 10$, $MP = x + 8$, and $NO = 2x + 1$. Solve for $x$.

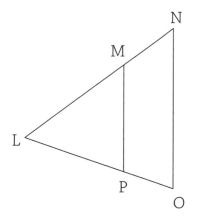

**5.** In the following diagram of right triangle *LMN*, altitude $\overline{MO}$ is drawn to hypotenuse $\overline{LN}$. If *LO* = 2 and *LN* = 20, what is the length of $\overline{MO}$?

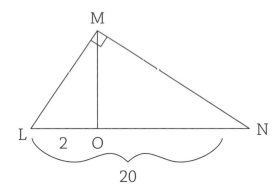

**6.** The accompanying diagram represents the location of four friends' homes along a right triangle. The distance from Jonathan's house to Zachary's house is 4 miles less than the distance from Jonathan's house to Andrew's house. The distance from Jonathan's house to Brandon's house is twice the distance from Jonathan's house to Andrew's house. How far away is Zachary's house from Brandon's house?

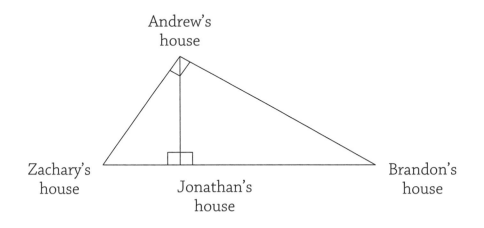

7. In the accompanying diagram of right triangle *ABC*, altitude $\overline{BD}$ is drawn to hypotenuse $\overline{AC}$. Solve for *x*.

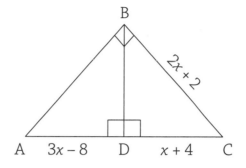

## EXERCISE 5-3

*Complete the following proofs using the given information.*

1. 

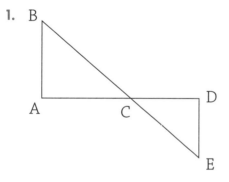

Given: $\overline{BE}$ intersects $\overline{AD}$ at *C*.

$\overline{BA} \perp \overline{AD}, \overline{ED} \perp \overline{AD}$

Prove: $\triangle BAC \sim \triangle EDC$

**2.**

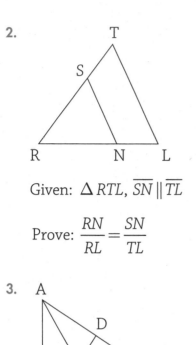

Given: $\triangle RTL$, $\overline{SN} \parallel \overline{TL}$

Prove: $\dfrac{RN}{RL} = \dfrac{SN}{TL}$

**3.**

Given: $\triangle ABC$ is isosceles with vertex at $\angle C$, $\overline{AE} \perp \overline{CB}$, $\overline{CD} \perp \overline{AB}$.

Prove: $BE \times CD = AD \times AE$

Flashcard
App

# 6 Getting to Know Right Triangles

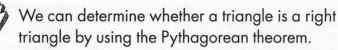

## MUST KNOW

 A triangle with an angle measuring 90° is a right triangle. The side opposite this angle is called the *hypotenuse*.

 We can determine whether a triangle is a right triangle by using the Pythagorean theorem.

 The sine, cosine, and tangent ratios are used to solve for the sides and angles of a right triangle.

eometry studies many different figures, right triangles being one of the most important.

In this chapter, we will learn different applications of the Pythagorean theorem and the trigonometric ratios sine, cosine, and tangent. We will also get to see the interesting relationships between the sides of the special 45°, 45°, 90°, and 30°, 60°, 90° right triangles.

## The Pythagorean Theorem

A right triangle contains three sides and three angles, one of which must be 90°. The largest side of the right triangle, the side opposite the 90° angle, is called the **hypotenuse**, whereas the two smaller sides are called **legs**. The Pythagorean theorem states that in a right triangle the sum of the squares of the two legs is equal to the square of the hypotenuse.

The **Pythagorean theorem** is $a^2 + b^2 = c^2$, where $a$ and $b$ represent the legs of the right triangle and $c$ represents the hypotenuse.

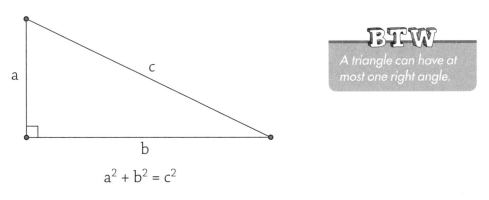

$$a^2 + b^2 = c^2$$

**BTW**
A triangle can have at most one right angle.

**IRL** The Pythagorean theorem is named after Greek mathematician and philosopher Pythagoras. He was born on the island of Samos, Greece, in 569 BC. While some knowledge of the theorem existed before Pythagoras, it is believed that he played a major role in the first proof of the theorem.

The Pythagorean theorem gives us the ability to solve for the missing side of a right triangle if we know the lengths of the remaining two sides. Let's practice this theorem by solving for the missing side in the following example.

▶ In the following diagram of right triangle $ABC$, $AC = 4$, and $BC = 3$. Find the length of $\overline{AB}$.

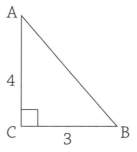

▶ $\overline{AC}$ and $\overline{BC}$ are legs of the right triangle, and $\overline{AB}$ is the hypotenuse of the right triangle because it is opposite the right angle. We can substitute the information into the Pythagorean theorem to solve for the length of $\overline{AB}$.

$$a^2 + b^2 = c^2 \quad \text{or} \quad leg^2 + leg^2 = hypotenuse^2$$
$$3^2 + 4^2 = AB^2$$
$$9 + 16 = AB^2$$
$$25 = AB^2$$

▶ To solve for $AB$, we will square root both sides. We know the length of a triangle cannot be negative, so we will take the **principal square root**. This means we will only have a positive solution for $AB$.

$$\sqrt{25} = \sqrt{AB^2}$$
$$5 = AB$$

The following example shows how to solve for the leg of a right triangle.

In the following diagram of right triangle $RST$, $ST = 5$, and $RT = 13$. Find the length of $\overline{RS}$.

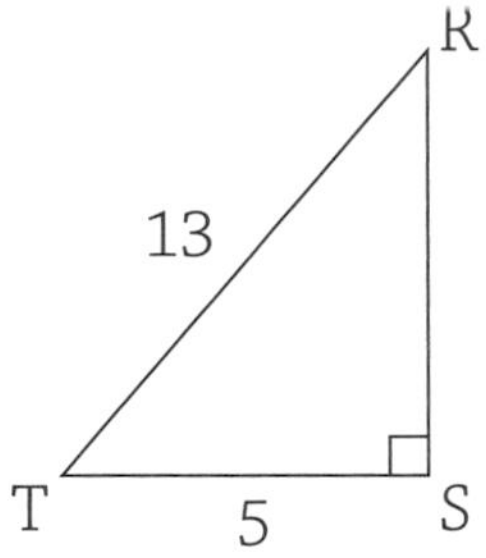

$\overline{ST}$ is a leg of the right triangle, and $\overline{RT}$ is the hypotenuse of the right triangle because it is opposite the right angle. We are looking to find the length of $\overline{RS}$, which is the other leg of the right triangle. We can substitute the information into the Pythagorean theorem to solve for the length of $\overline{RS}$.

$$a^2 + b^2 = c^2 \quad \text{or} \quad leg^2 + leg^2 = hypotenuse^2$$

$$5^2 + RS^2 = 13^2$$

$$25 + RS^2 = 169$$
$$-\,25 \qquad\quad -25$$
$$\overline{\phantom{xxxxxxxxxx}}$$
$$RS^2 = 144$$
$$\sqrt{RS^2} = \sqrt{144}$$
$$RS = 12$$

In addition to using the Pythagorean theorem to determine the lengths of missing sides of given right triangles, we can also use the Pythagorean theorem to determine whether a triangle is a right triangle if we are given the lengths of all three sides of the triangle. Can you use the Pythagorean theorem to check to see which of the two triangles below is a right triangle?

EXAMPLE

Triangle 1:

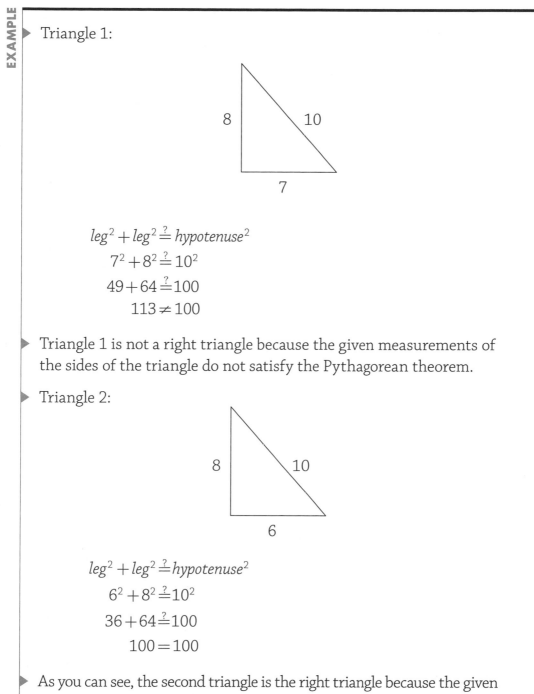

$$leg^2 + leg^2 \stackrel{?}{=} hypotenuse^2$$
$$7^2 + 8^2 \stackrel{?}{=} 10^2$$
$$49 + 64 \stackrel{?}{=} 100$$
$$113 \neq 100$$

Triangle 1 is not a right triangle because the given measurements of the sides of the triangle do not satisfy the Pythagorean theorem.

Triangle 2:

$$leg^2 + leg^2 \stackrel{?}{=} hypotenuse^2$$
$$6^2 + 8^2 \stackrel{?}{=} 10^2$$
$$36 + 64 \stackrel{?}{=} 100$$
$$100 = 100$$

As you can see, the second triangle is the right triangle because the given measurements of the sides of the triangle satisfy the Pythagorean theorem.

# Pythagorean Triples

Have you ever noticed that certain lengths of sides tend to show up frequently in right triangles? We call these **Pythagorean triples**. Pythagorean triples are integer solutions to the Pythagorean theorem. Knowing Pythagorean triples helps you solve for the missing sides of right triangles in a more efficient way. Some examples of Pythagorean triples are {3,4,5}, {5,12,13}, {8,15,17}, and {7,24,25}. You can see that each set of triples satisfies the Pythagorean theorem.

$$3^2 + 4^2 = 5^2$$
$$5^2 + 12^2 = 13^2$$
$$8^2 + 15^2 = 17^2$$
$$7^2 + 24^2 = 25^2$$

Each Pythagorean triple can give an infinite number of new triples. For example, if you double each side of the {3,4,5} triangle, you will get a new triple {6,8,10}. By tripling each side of the {3,4,5} triangle, you would get {9,12,15}.

**BTW**

*A primitive triple is a triple where the only common factor for the lengths of the three sides of the triangle is the number 1. There are 16 primitive Pythagorean triples with a hypotenuse measuring less than 100:*

*{3, 4, 5}*
*{5, 12, 13}*
*{8, 15, 17}*
*{7, 24, 25}*
*{20, 21, 29}*
*{12, 35, 37}*
*{9, 40, 41}*
*{28, 45, 53}*
*{11, 60, 61}*
*{33, 56, 65}*
*{16, 63, 65}*
*{48, 55, 73}*
*{36, 77, 85}*
*{13, 84, 85}*
*{39, 80, 89}*
*{65, 72, 97}*

# Special Right Triangles

The most common right triangles that we study in geometry are the 45°, 45°, 90° and 30°, 60°, 90° triangles. We call these **special right triangles** because their sides always follow a specific ratio. Recognizing the special right triangles and understanding their ratios will make calculations easier and more efficient.

A 45°, 45°, 90° triangle contains two equal angles, which classifies it as an isosceles triangle. Because this triangle also contains a right angle, it is called an **isosceles right triangle**. The following diagram shows the ratio of the sides of this special isosceles right triangle.

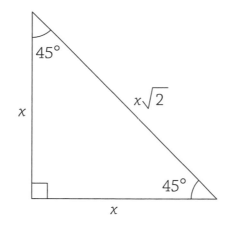

The ratio of the length of the sides are $1:1:\sqrt{2}$, or it can be labeled as $x: x: x\sqrt{2}$ with the two congruent sides, $x$, opposite the congruent 45° angles, and the hypotenuse, $x\sqrt{2}$, opposite the 90° angle. This means that in an isosceles right triangle the length of the hypotenuse is always $\sqrt{2}$ times the length of either congruent leg.

**BTW**

*If two angles of a triangle are congruent, then the sides opposite those angles are congruent.*

The following example shows how to use the ratios for the sides of a 45°, 45°, 90° triangle to find the length of the hypotenuse of the triangle.

**EXAMPLE**

▶ The following diagram shows a 45°, 45°, 90° triangle with both legs measuring three units. Find the length of the hypotenuse.

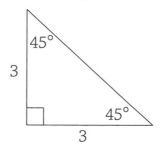

▶ Let's fill in the ratio for the sides of a 45°, 45°, 90° triangle into our diagram.

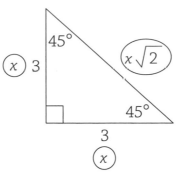

▶ Let $x =$ the measure of the legs. In this diagram, the legs measure three units, which means $x = 3$.

▶ Let $x\sqrt{2} =$ the measure of the hypotenuse. We already determined that $x = 3$; therefore, the hypotenuse must be $3\sqrt{2}$.

## EXTRA HELP

We can use the Pythagorean theorem to check the solution.

$$a^2 + b^2 = c^2$$
$$3^2 + 3^2 = c^2$$
$$18 = c^2$$
$$\sqrt{18} = c$$
$$3\sqrt{2} = c$$

Remember that the lengths of the sides of your triangles are not always going to be integers. The following example shows how to use the ratios for the sides of a 45°, 45°, 90° triangle to find the length of the hypotenuse of the triangle when the legs have irrational length.

EXAMPLE

▶ The following diagram shows a 45°, 45°, 90° triangle with both legs measuring $5\sqrt{2}$ units. Find the length of the hypotenuse.

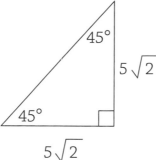

▶ Let's fill in the ratio for the sides of a 45°, 45°, 90° triangle into our diagram.

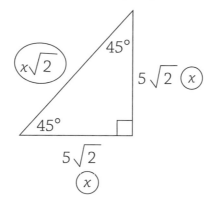

▶ Let $x$ = the measure of the legs. In this diagram, $x = 5\sqrt{2}$.

▶ Let $x\sqrt{2}$ = the measure of the hypotenuse. We already determined $x = 5\sqrt{2}$; therefore, the hypotenuse must be:

$$5\sqrt{2} \cdot \sqrt{2} = 5\sqrt{4}$$
$$= 5(2)$$
$$= 10$$

Now let's see how our solutions will change when we are given the length of the hypotenuse of the 45°, 45°, 90° triangle and need to find the length of a leg. Because a 45°, 45°, 90° triangle is an isosceles triangle, both legs are equal in length.

▶ The following diagram shows a 45°, 45°, 90° triangle with a hypotenuse measuring $10\sqrt{2}$. Find the length of the legs of the triangle.

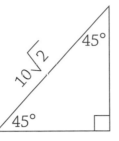

▶ Let's fill in the ratio for the sides of a 45°, 45°, 90° triangle into our diagram.

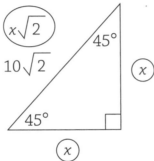

▶ Let $x =$ the measure of the legs.

▶ Let $x\sqrt{2} =$ the measure of the hypotenuse.

▶ Because we know the length of the hypotenuse is $10\sqrt{2}$, we can set up $x\sqrt{2} = 10\sqrt{2}$. To solve for $x$, we need to divide both sides by $\sqrt{2}$:

$$\frac{x\cancel{\sqrt{2}}}{\cancel{\sqrt{2}}} = \frac{10\cancel{\sqrt{2}}}{\cancel{\sqrt{2}}}$$

$$x = 10$$

Don't forget that the length of the legs can also be irrational. The following example shows how to use the ratios for the sides of a 45°, 45°, 90° triangle to find the length of the legs of the triangle when the length of the hypotenuse is an integer. In this situation, we will have to rationalize the denominator at the end of our work.

EXAMPLE

▶ The following diagram shows a 45°, 45°, 90° triangle with a hypotenuse measuring 8. Find the length of the legs of the triangle.

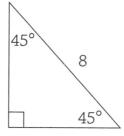

▶ Let's fill in the ratio for the sides of a 45°, 45°, 90° triangle into our diagram.

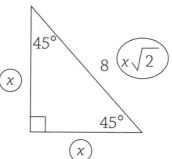

▶ Let $x$ = the measure of the legs.

▶ Let $x\sqrt{2}$ = the measure of the hypotenuse.

▶ Because we already know the hypotenuse has a length of 8, we are able to set up $x\sqrt{2} = 8$. Divide both sides by $\sqrt{2}$:

$$\frac{x\cancel{\sqrt{2}}}{\cancel{\sqrt{2}}} = \frac{8}{\sqrt{2}}$$

$$x = \frac{8}{\sqrt{2}}$$

▶ Rationalize the denominator by multiplying the numerator and denominator by $\sqrt{2}$:

$$x = \frac{8\sqrt{2}}{\sqrt{2}\sqrt{2}}$$

$$x = \frac{8\sqrt{2}}{2}$$

▶ Therefore, the length of the leg is $4\sqrt{2}$.

**BTW**

To make the denominator of a fraction rational, you must multiply the numerator and denominator of the fraction by the radical in the denominator.

Now that we understand the ratios of a 45°, 45°, 90° triangle, let's learn about the other special right triangle called the **30°, 60°, 90° triangle**. Being that a 30°, 60°, 90° triangle does not have any equal angles, we know that the sides are all different in length. The lengths of the sides of the triangle are in the ratio $1:\sqrt{3}:2$ and can be labeled as $x:x\sqrt{3}:2x$. As you can see in the following diagram, the side opposite the 30° angle (the smallest side) is represented by $x$, the side opposite the 60° angle is represented by $x\sqrt{3}$, and the side opposite the 90° angle (the hypotenuse and largest side) is represented by $2x$.

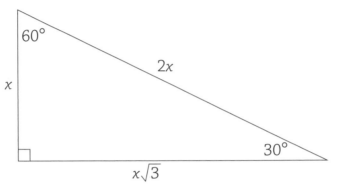

The following example shows how to use the ratios of a 30°, 60°, 90° triangle to solve for the lengths of two sides of the triangle when given the length of the side opposite the 30° angle.

The following diagram shows a 30°, 60°, 90° triangle with the smallest side measuring 7. Find the lengths of the other two sides of the triangle.

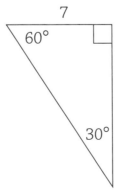

▶ Let's fill in the ratio for the sides of a 30°, 60°, 90° triangle into our diagram.

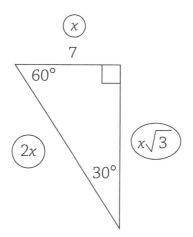

▶ Let $x$ = the length of the side opposite the 30° angle.

▶ Let $x\sqrt{3}$ = the length of the side opposite the 60° angle.

▶ Let $2x$ = the length of the hypotenuse.

▶ We know the length of the side opposite the 30° angle is 7; therefore, $x = 7$. We will be using this value of $x$ to solve for the lengths of the other two sides. The hypotenuse is represented by $2x$, which means the length of the hypotenuse is $2(7) = 14$. The side opposite the 60° angle is represented by $x\sqrt{3}$, which means the length of that side is $7\sqrt{3}$.

In the next example, we will use the ratios of a 30°, 60°, 90° triangle to find the lengths of two sides of the triangle when we are given the length of the hypotenuse.

▶ The following diagram shows a 30°, 60°, 90° triangle with the hypotenuse measuring 10. Find the lengths of the other two sides of the triangle.

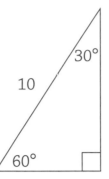

▶ Let's fill in the ratio for the sides of a 30°, 60°, 90° triangle in our diagram.

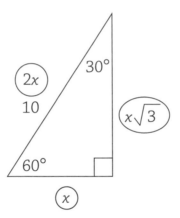

▶ Let $x$ = the measure of the side opposite the 30° angle.

▶ Let $x\sqrt{3}$ = the measure of the side opposite the 60° angle.

▶ Let $2x$ = the measure of the hypotenuse.

▶ We were given that the length of the hypotenuse is 10. Because the hypotenuse is represented as $2x$, the equation $2x = 10$ gives us the ability to solve for $x$. This means $x$ equals 5. We will substitute this value of $x$ into the ratios of the sides of a 30°, 60°, 90° triangle to solve for the remaining sides.

▶ The side opposite the 30° angle is represented by $x$; therefore, it is equal to 5. The side opposite the 60° angle is represented by $x\sqrt{3}$; therefore, it is equal to $5\sqrt{3}$.

Finding the lengths of the sides of a 30°, 60°, 90° triangle when given the length of the side opposite the 60° angle can prove to be a more challenging question. We will address this in the next example.

EXAMPLE

▶ The following diagram shows a 30°, 60°, 90° triangle with the side opposite the 60° angle measuring $9\sqrt{3}$. Find the lengths of the other two sides of the triangle.

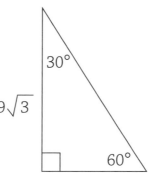

▶ Let's fill in the ratio for the sides of a 30°, 60°, 90° triangle in our diagram.

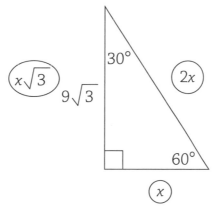

▶ Let $x =$ the length of the side opposite the 30° angle.

▶ Let $x\sqrt{3} =$ the length of the side opposite the 60° angle.

▶ Let $2x =$ the length of the hypotenuse.

▶ We know that the side opposite the 60° angle is $9\sqrt{3}$. Because the ratio tells us that the side opposite the 60° angle is $x\sqrt{3}$, we can set up the equation $x\sqrt{3} = 9\sqrt{3}$ to solve for $x$. We divide $\sqrt{3}$ from both sides:

$$\frac{x\sqrt{3}}{\sqrt{3}} = \frac{9\sqrt{3}}{\sqrt{3}} \text{ and get } x = 9.$$

▶ The length of the side opposite the 30° angle is represented by $x$, which means the side opposite the 30° angle is 9.

> **BTW**
>
> In a 30°, 60°, 90° triangle, dividing the side opposite the 60° angle by $\sqrt{3}$ will result in the length of the side opposite the 30° angle.

▶ The ratio shows the hypotenuse is represented by $2x$. When we substitute in the value of $x$, we get $2(9) = 18$. Therefore, the length of the hypotenuse is 18.

## Right Triangle Trigonometry

Now that we are getting more comfortable with right triangles, it is extremely important that we learn about the trigonometric functions **sin (sine)**, **cos (cosine)**, and **tan (tangent)**. Each function corresponds to a ratio of the sides of a right triangle.

We call these three sides the **hypotenuse**, **adjacent**, and **opposite** sides. The hypotenuse is always found directly across from the right angle. The adjacent side is the side next to the angle in question, and the opposite side is the side opposite the angle in question.

The following diagram shows two labeled right triangles. The first triangle is labeled based on angle $A$, and the second triangle is labeled based on angle $B$.

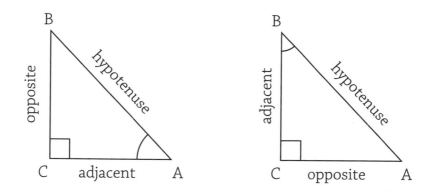

We call sine, cosine, and tangent trigonometric ratios because they are each equal to the ratio of two of the sides in the right triangle:

$$\sin\theta = \frac{\text{opposite}}{\text{hypotenuse}}$$

$$\cos\theta = \frac{\text{adjacent}}{\text{hypotenuse}}$$

$$\tan\theta = \frac{\text{opposite}}{\text{adjacent}}$$

**BTW**

*The greek letter $\theta$ is used to represent an angle.*

 **IRL** *Soh Cah Toa* **is one of the most popular ways students remember the trig ratios. The capital letter stands for the trig function, and the lowercase letters stand for the two necessary sides.**

Let's get some practice setting up the trig ratios. Using the preceding diagrams, we can see that:

$$\sin A = \frac{BC}{AB} \quad \cos A = \frac{AC}{AB} \quad \tan A = \frac{BC}{AC}$$

$$\sin B = \frac{AC}{AB} \quad \cos B = \frac{BC}{AB} \quad \tan B = \frac{AC}{BC}$$

We can use these trigonometric ratios not only to find the length of a side of a triangle but also to find an angle of the triangle. To be able to find the length of a side of a triangle using the trigonometric ratios, you must know the length of at least one side of the right triangle and know the measure of one of the angles in the triangle that is not 90°.

The following example uses trigonometric ratios to solve for the hypotenuse of a right triangle.

**EXAMPLE**

▶ Solve for $x$ in the following diagram. Round your answer to the nearest hundredth.

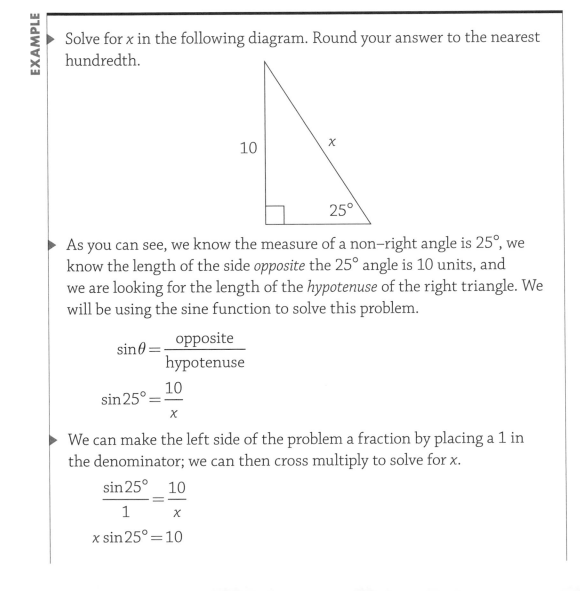

▶ As you can see, we know the measure of a non–right angle is 25°, we know the length of the side *opposite* the 25° angle is 10 units, and we are looking for the length of the *hypotenuse* of the right triangle. We will be using the sine function to solve this problem.

$$\sin\theta = \frac{\text{opposite}}{\text{hypotenuse}}$$

$$\sin 25° = \frac{10}{x}$$

▶ We can make the left side of the problem a fraction by placing a 1 in the denominator; we can then cross multiply to solve for $x$.

$$\frac{\sin 25°}{1} = \frac{10}{x}$$

$$x \sin 25° = 10$$

▶ We can now divide both sides by sin 25° to solve for x.

$$\frac{x\sin 25°}{\sin 25°}=\frac{10}{\sin 25°}$$

$$x = 23.66$$

Let's try a problem that requires a different trigonometric function to solve for the length of a side of the triangle.

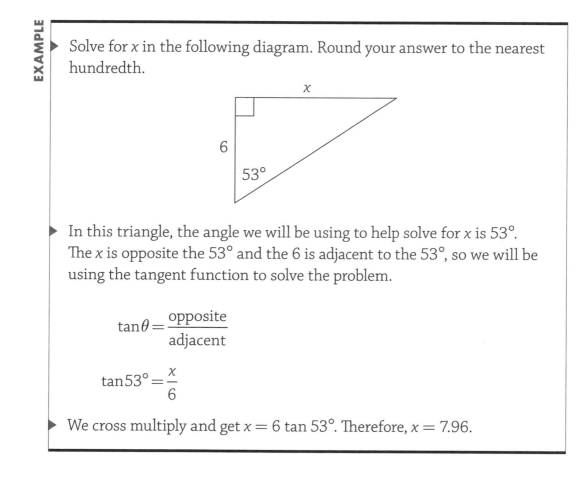

**EXAMPLE**

▶ Solve for x in the following diagram. Round your answer to the nearest hundredth.

▶ In this triangle, the angle we will be using to help solve for x is 53°. The x is opposite the 53° and the 6 is adjacent to the 53°, so we will be using the tangent function to solve the problem.

$$\tan\theta=\frac{\text{opposite}}{\text{adjacent}}$$

$$\tan 53°=\frac{x}{6}$$

▶ We cross multiply and get x = 6 tan 53°. Therefore, x = 7.96.

Solving for an angle of a right triangle is a little different than solving for a side. To solve for an angle, we will need to use the inverse of the trigonometric functions: $\sin^{-1}$, $\cos^{-1}$, and $\tan^{-1}$. The inverse trig functions give us the ability to use the answer to the question to find the angle.

The following example uses the inverse trigonometric function to solve for the angle of a right triangle.

**EXAMPLE**

▶ Solve for $x$ in the following diagram. Round your answer to the nearest hundredth of a degree.

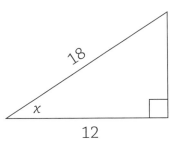

▶ The side measuring 12 is adjacent to the angle in question, and the side measuring 18 is the hypotenuse of the triangle, which means we will be using the cosine function to solve the problem.

$$\cos\theta = \frac{\text{adjacent}}{\text{hypotenuse}}$$

$$\cos x = \frac{12}{18}$$

▶ Cross multiplying will not help us in this situation because the $x$ is not one of the sides. Therefore, we will set up our inverse trig function.

$$x = \cos^{-1}\left(\frac{12}{18}\right)$$

$$= 48.19°$$

## Word Problems

How could we end a chapter on trigonometry without looking at word problems? The key to making word problems easier is to draw a diagram. Let's first look at a basic trigonometric word problem. We will then introduce some more challenging vocabulary.

▶ A 12-foot ladder is leaning against a house, making a 62° angle with the ground. How far, to the nearest tenth of a foot, is the base of the ladder from the house?

▶ Our first step will always be to draw a diagram.

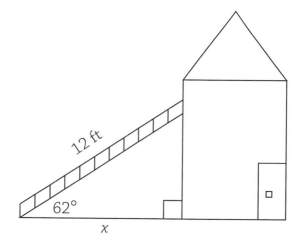

▶ As you can see in the diagram, the ladder is the hypotenuse of a right triangle and we are looking for the length of the side adjacent to the 62° angle. We will be using the cosine function to solve for the missing side.

$$\cos\theta = \frac{\text{adjacent}}{\text{hypotenuse}}$$

$$\cos 62° = \frac{x}{12}$$

> We can cross multiply to solve for $x$:
>
> $$x = 12 \cos 62°$$
> $$= 5.6 \text{ feet}$$

Two phrases appear frequently in right-triangle word problems: **angle of elevation** and **angle of depression**. *Angle of elevation* refers to the angle formed from the horizontal segment representing the ground, up toward the segment representing the line of sight. *Angle of depression* refers to the angle formed from the horizontal segment at the highest point (normally not drawn) down toward the segment representing the line of sight.

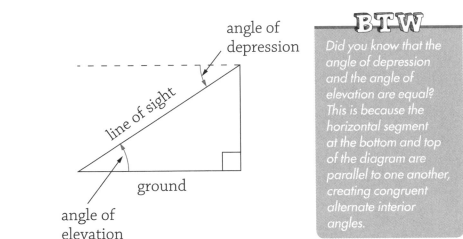

**BTW**

*Did you know that the angle of depression and the angle of elevation are equal? This is because the horizontal segment at the bottom and top of the diagram are parallel to one another, creating congruent alternate interior angles.*

It is time for us to put all our knowledge of trigonometry to the test. In our first example, we will need to use trigonometry to find the length of a side of a triangle.

EXAMPLE

▶ Michael is standing on the ground 200 feet from the base of a building. The angle of elevation from his location to the top of the building is found to be 10°. Find the height of the building to the *nearest foot*.

▶ Our first step will always be to draw a diagram:

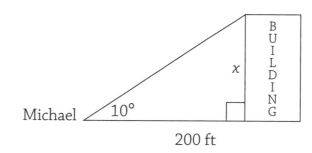

▶ The side measuring 200 feet is adjacent to the 10° angle, and we are looking for the length of the side opposite the 10° angle. We will be using the tangent function.

$$\tan\theta = \frac{\text{opposite}}{\text{adjacent}}$$

$$\tan 10° = \frac{x}{200}$$

▶ We will now cross multiply to solve for $x$:

$$x = 200 \tan 10°$$

$$= 35 \text{ feet}$$

Our last example will use trigonometry to find the measure of an angle of a triangle.

▶ A man standing 100 feet high in a lighthouse catches sight of a boat in the water. If the boat is 800 feet from the base of the lighthouse, find the angle of depression from the man in the lighthouse to the boat to the *nearest tenth of a degree.*

▶ Let's begin by drawing a diagram. Because we already learned that the angle of depression and the angle of elevation are equal, our diagram should look like this:

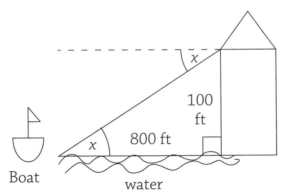

▶ We know the side measuring 100 feet is opposite angle $x$, and the side measuring 800 feet is adjacent to angle $x$, so we will be using the tangent function.

$$\tan\theta = \frac{\text{opposite}}{\text{adjacent}}$$

$$\tan x = \frac{100}{800}$$

▶ We already know the answer and are looking for the angle, so we will be using the inverse of tangent.

$$x = \tan^{-1}\left(\frac{100}{800}\right)$$

$$= 7.1°$$

## EXERCISES

### EXERCISES 6-1

*Follow the instructions for each question.*

1. Solve for the missing side in the following right triangle.

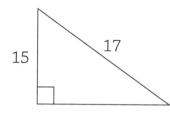

2. Determine if the following triangle is a right triangle.

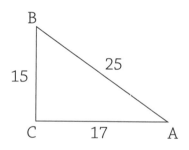

3. Use the ratios of a 45°, 45°, 90° triangle to solve for the length of $\overline{AB}$.

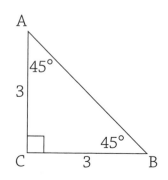

**4.** Use the ratios of a 45°, 45°, 90° triangle to solve for the length of $\overline{AC}$.

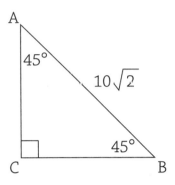

**5.** Use the ratios of a 45°, 45°, 90° triangle to solve for the length of $\overline{BC}$.

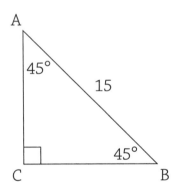

**6.** Use the ratios of a 30°, 60°, 90° triangle to solve for the length of $\overline{AB}$.

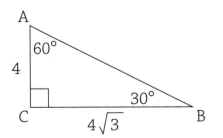

7. Use the ratios of a 30°, 60°, 90° triangle to solve for the length of $\overline{BC}$.

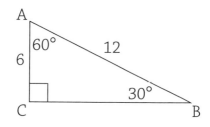

8. Use the ratios of a 30°, 60°, 90° triangle to solve for the length of $\overline{AC}$ and $\overline{BC}$.

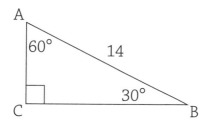

9. Use the ratios of a 30°, 60°, 90° triangle to solve for the length of $\overline{AC}$ and $\overline{AB}$.

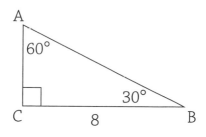

10. Find $x$ to the *nearest tenth* using trigonometric ratios.

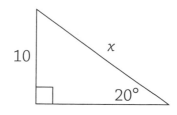

**11.** Find *x* to the *nearest hundredth* using trigonometric ratios.

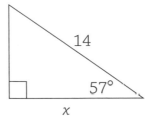

**12.** Find *x* to the *nearest tenth of a degree* using trigonometric ratios.

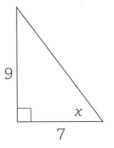

**13.** Find *x* to the *nearest hundredth of a degree* using trigonometric ratios.

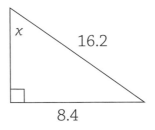

**14.** Gordon is standing 40 feet from the base of a building. The angle of elevation from his location to the top of the building is 68°. Find the height of the building to the *nearest foot*.

**15.** Sherry, who is 5.5 feet tall, is standing on the roof of a 50-foot building. She looks down at a spot on the ground 20 feet from the base of the building. Find the angle of depression to the *nearest tenth of a degree*.

# Parallel Lines

# MUST ⚡ KNOW

 When cut by a line called a *transversal*, parallel lines form four kinds of angles with unique properties: alternate interior angles, corresponding angles, alternate exterior angles, and same-side interior angles.

These pairs of angles are either congruent—they are equal—or supplementary—their sum equals 180°.

Alternate interior angles, corresponding angles, and alternate exterior angles are congruent.

Same-side interior angles and same-side exterior angles are supplementary.

hat comes to mind when you think of parallel lines? Many train tracks run parallel to one another. The edges of rooftops, sides of tables, and yard lines on a football field are also examples of parallel lines. In geometry, parallel lines are defined as two lines on a plane that never intersect and are always the same distance apart. Any line that cuts through parallel lines is called a **transversal**.

> **IRL** Part of the international boundary between Canada and the United States is the 49th parallel north. The term *parallel* is the imaginary line representing the degrees of latitude that goes around Earth and is parallel to the plane of the equator.

## Alternate Interior Angles

Two parallel lines that are cut by a transversal form **alternate interior angles** that are congruent. The following diagram shows two pairs of parallel lines cut by a transversal. The angles that are formed on opposite sides of the transversal and inside the parallel lines are the alternate interior angles. In this diagram, the pairs of alternate interior angles that are congruent are $\angle 1 \cong \angle 2$ and $\angle 3 \cong \angle 4$.

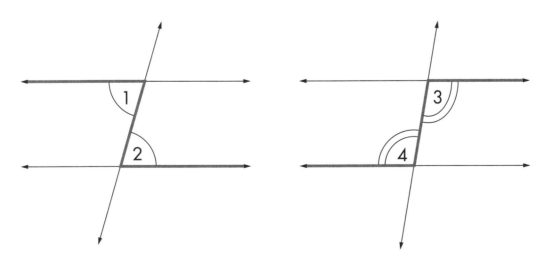

## EXTRA HELP

A trick to help identify the alternate interior angles is to highlight a *Z* in the diagram by tracing the two parallel lines and the transversal. The angles inside the *Z* will be a pair of alternate interior angles (∠1 and ∠2). The *Z* can also be highlighted backward. The backward Z will help you identify the other pair of alternate interior angles (∠3 and ∠4).

Let's look at an example in which recognizing congruent alternate interior angles helps solve an algebraic problem.

**EXAMPLE**

In the following diagram, $\overline{AB}$ is parallel to $\overline{CD}$. If $m\angle AEF = 2x + 1$ and $m\angle DFE = x + 50$, what is $m\angle DFE$?

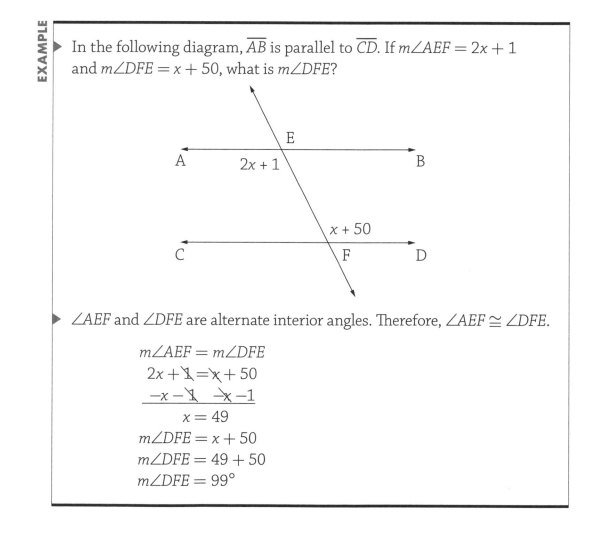

∠*AEF* and ∠*DFE* are alternate interior angles. Therefore, ∠*AEF* ≅ ∠*DFE*.

$$m\angle AEF = m\angle DFE$$
$$2x + 1 = x + 50$$
$$\underline{-x - 1 \quad -x - 1}$$
$$x = 49$$
$$m\angle DFE = x + 50$$
$$m\angle DFE = 49 + 50$$
$$m\angle DFE = 99°$$

In the following example, identifying the alternate interior angles will help prove that two triangles are congruent in a geometric proof.

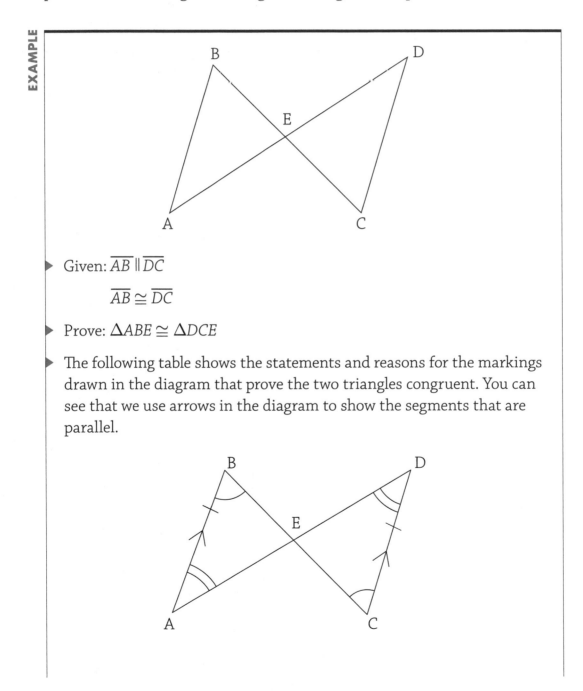

▶ Given: $\overline{AB} \parallel \overline{DC}$

   $\overline{AB} \cong \overline{DC}$

▶ Prove: $\triangle ABE \cong \triangle DCE$

▶ The following table shows the statements and reasons for the markings drawn in the diagram that prove the two triangles congruent. You can see that we use arrows in the diagram to show the segments that are parallel.

| Statement | Reason |
|---|---|
| 1. $\overline{AB} \parallel \overline{DC}$ | 1. Given |
| 2. $\angle ABE$ and $\angle DCE$, $\angle BAE$ and $\angle CDE$ are alternate interior angles. | 2. Parallel lines cut by a transversal form alternate interior angles. |
| 3. $\angle ABE \cong \angle DCE$ $\angle BAE \cong \angle CDE$ | 3. Alternate interior angles are congruent. |
| 4. $\overline{AB} \cong \overline{DC}$ | 4. Given |
| 5. $\triangle ABE \cong \triangle DCE$ | 5. ASA $\cong$ ASA |

## Corresponding Angles

Two parallel lines that are cut by a transversal form **corresponding angles** that are congruent. The following diagram shows two pairs of parallel lines cut by a transversal. Corresponding angles are the angles formed on the same side of the transversal with both either above or below the parallel lines. In this diagram, the pairs of corresponding angles that are congruent are $\angle 1 \cong \angle 2$, $\angle a \cong \angle b$, $\angle 3 \cong \angle 4$, and $\angle c \cong \angle d$.

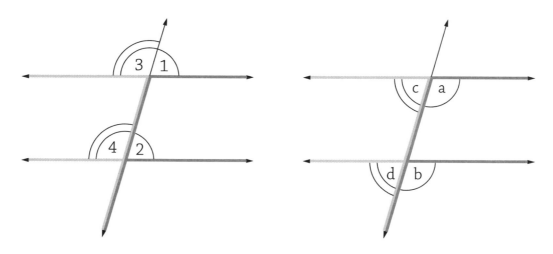

One way to help identify the corresponding angles is to highlight an "F" in the diagram by tracing the two parallel lines and the transversal. The F can also be highlighted backward. The angles that are both above the F are corresponding angles (∠1, ∠2 and ∠3, ∠4). The angles that are both below the F are another set of corresponding angles (∠a, ∠b and ∠c, ∠d).

Let's look at an example where recognizing congruent corresponding angles helps solve an algebraic problem.

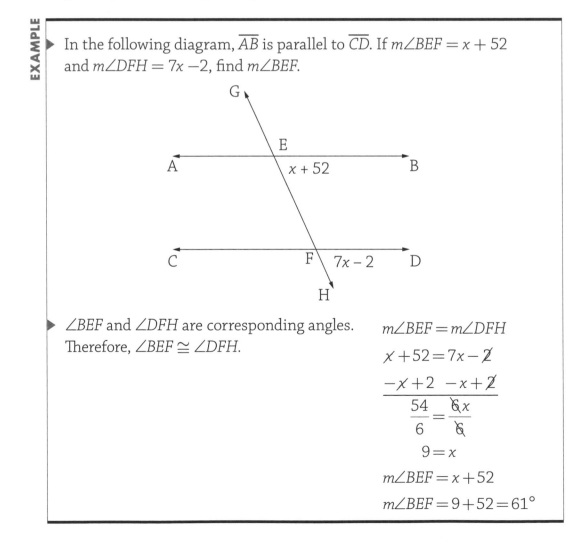

In the following diagram, $\overline{AB}$ is parallel to $\overline{CD}$. If $m\angle BEF = x + 52$ and $m\angle DFH = 7x - 2$, find $m\angle BEF$.

∠BEF and ∠DFH are corresponding angles. Therefore, ∠BEF ≅ ∠DFH.

$$m\angle BEF = m\angle DFH$$
$$x + 52 = 7x - 2$$
$$-x + 2 \quad -x + 2$$
$$\frac{54}{6} = \frac{6x}{6}$$
$$9 = x$$
$$m\angle BEF = x + 52$$
$$m\angle BEF = 9 + 52 = 61°$$

In the following example, identifying corresponding angles will help prove that two triangles are congruent in a geometric proof.

EXAMPLE

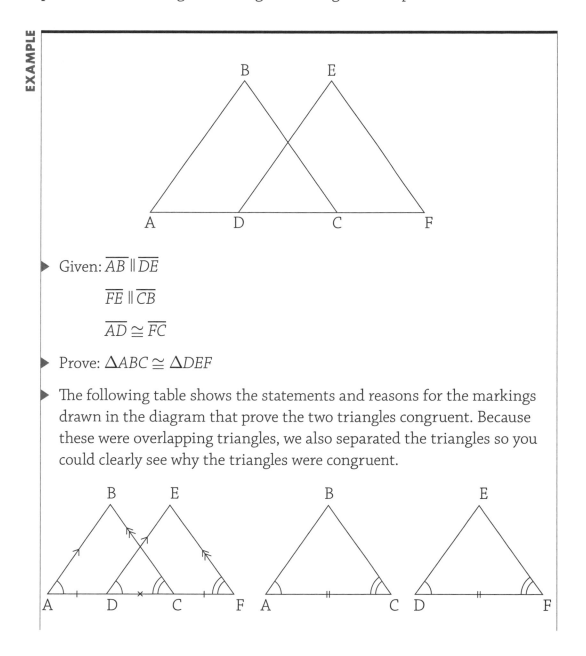

▶ Given: $\overline{AB} \parallel \overline{DE}$

  $\overline{FE} \parallel \overline{CB}$

  $\overline{AD} \cong \overline{FC}$

▶ Prove: $\triangle ABC \cong \triangle DEF$

▶ The following table shows the statements and reasons for the markings drawn in the diagram that prove the two triangles congruent. Because these were overlapping triangles, we also separated the triangles so you could clearly see why the triangles were congruent.

| Statement | Reason |
|---|---|
| 1. $\overline{AB} \parallel \overline{DE}$ <br> $\overline{FE} \parallel \overline{CB}$ | 1. Given |
| 2. ∠BAC and ∠EDC, ∠EFD and ∠BCA are corresponding angles. | 2. Parallel lines cut by a transversal form corresponding angles. |
| 3. ∠BAC ≅ ∠EDC <br> ∠EFD ≅ ∠BCA | 3. Corresponding angles are congruent. |
| 4. $\overline{AD} \cong \overline{FC}$ | 4. Given |
| 5. $\overline{DC} \cong \overline{DC}$ | 5. Reflexive property |
| 6. $\overline{AC} \cong \overline{FD}$ | 6. Addition postulate |
| 7. △ABC ≅ △DEF | 7. ASA ≅ ASA |

## Alternate Exterior, Same-Side Interior, and Same-Side Exterior Angles

Two parallel lines that are cut by a transversal form **alternate exterior angles** that are congruent. The following diagram shows two pairs of parallel lines cut by a transversal. Alternate exterior angles are angles that are formed on opposite sides of the transversal and outside the parallel lines. In this diagram, the pairs of alternate exterior angles that are congruent are ∠a ≅ ∠b and ∠1 ≅ ∠2.

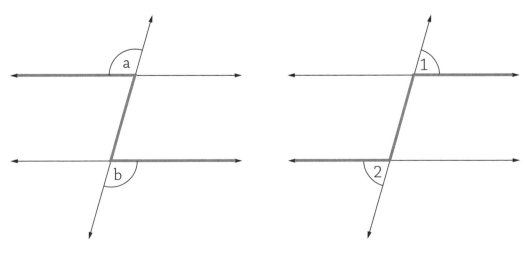

## EXTRA HELP

A tip to help identify the alternate exterior angles is to trace the two parallel lines and the transversal to look for the Z. The angles outside the Z are alternate exterior angles.

Now let's look at an example where recognizing congruent alternate exterior angles helps solve an algebraic problem.

**EXAMPLE**

▶ In the following diagram, $\overline{AB}$ is parallel to $\overline{CD}$. If $m\angle GEA = 8x + 10$ and $m\angle HFD = 2x + 34$, find $m\angle GEA$.

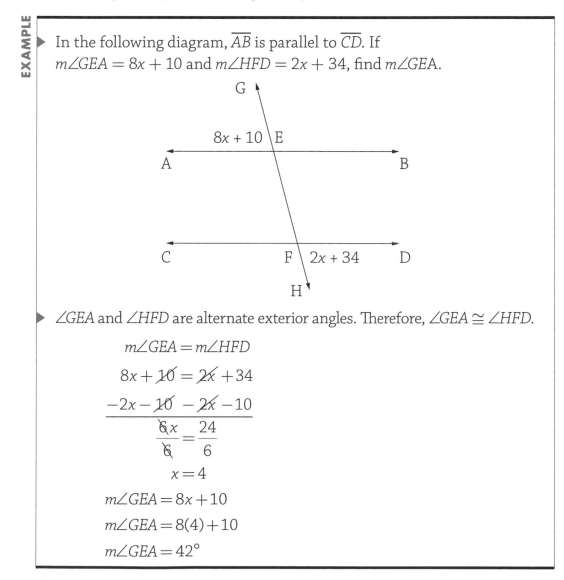

▶ $\angle GEA$ and $\angle HFD$ are alternate exterior angles. Therefore, $\angle GEA \cong \angle HFD$.

$$m\angle GEA = m\angle HFD$$

$$8x + \cancel{10} = \cancel{2x} + 34$$

$$\frac{-2x - \cancel{10} \quad -\cancel{2x} - 10}{\cancel{6}x = \frac{24}{6}}$$

$$\frac{\cancel{6}x}{\cancel{6}} = \frac{24}{6}$$

$$x = 4$$

$$m\angle GEA = 8x + 10$$

$$m\angle GEA = 8(4) + 10$$

$$m\angle GEA = 42°$$

Two parallel lines that are cut by a transversal form **same-side interior angles** that are supplementary and **same-side exterior angles** that are supplementary. The following diagram shows two parallel lines cut by a transversal. Same-side interior angles are on the same side of the transversal and inside the parallel lines.

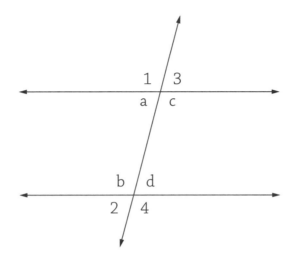

In this diagram, the pairs of same-side interior angles that are supplementary are $m\angle a + m\angle b = 180°$ and $m\angle c + m\angle d = 180°$. Same-side exterior angles are on the same side of the transversal and outside the parallel lines. In this diagram, the pairs of same-side exterior angles that are supplementary are $m\angle 1 + m\angle 2 = 180°$ and $m\angle 3 + m\angle 4 = 180°$.

## EXTRA HELP

We know that $\angle a \cong \angle 2$ because corresponding angles are congruent, and $\angle 2 + \angle b = 180$ since two angles that form a linear pair are supplementary. Therefore, by substituting $\angle a$ for $\angle 2$, $\angle a + \angle b = 180$.

Let's look at an example where recognizing supplementary same-side interior angles helps solve an algebraic problem.

EXAMPLE

▶ In the following diagram, $\overleftrightarrow{AB}$ is parallel to $\overleftrightarrow{CD}$. If $m\angle AEF = 2x + 60$ and $m\angle EFC = 4x$, find $m\angle AEF$.

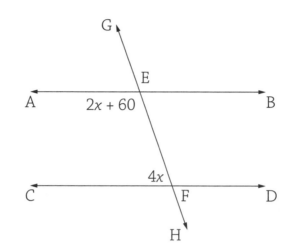

▶ $\angle AEF$ and $\angle EFC$ are same-side interior angles, which means they are supplementary angles. Supplementary angles have a sum of 180°.

$$m\angle AEF + m\angle EFC = 180°$$
$$2x + 60 + 4x = 180$$
$$6x + 60 = 180$$
$$\underline{-60 \quad -60}$$
$$\frac{6x}{6} = \frac{120}{6}$$
$$x = 20$$
$$m\angle AEF = 2x + 60$$
$$m\angle AEF = 2(20) + 60$$
$$m\angle AEF = 100°$$

## Auxiliary Lines

An **auxiliary line** is a line that can be drawn parallel to a given line passing through a given point. It gives us the ability to find the value of an angle by creating alternate interior, corresponding, or same-side interior angles.

The following example shows how to draw an auxiliary line to solve for the measure of an angle.

▶ In the following diagram, $\overline{AB} \parallel \overline{DC}$, $m\angle ABE = 45°$, and $m\angle ECD = 52°$. Find $m\angle BEC$.

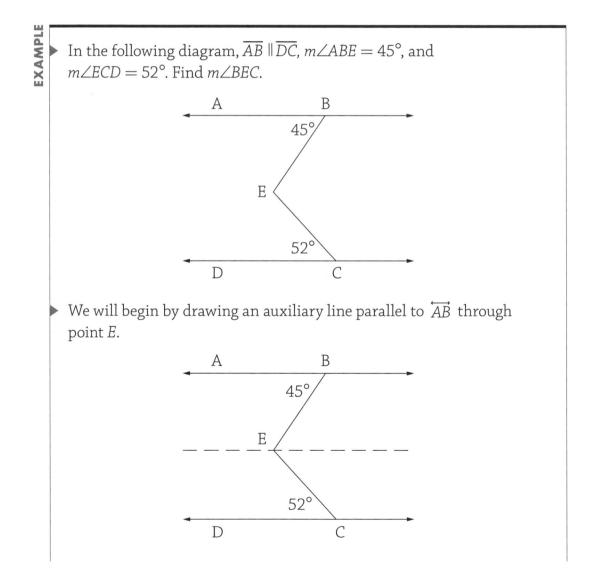

▶ We will begin by drawing an auxiliary line parallel to $\overleftrightarrow{AB}$ through point $E$.

▶ We can now look for congruent alternate interior angles and place their measurements in the diagram.

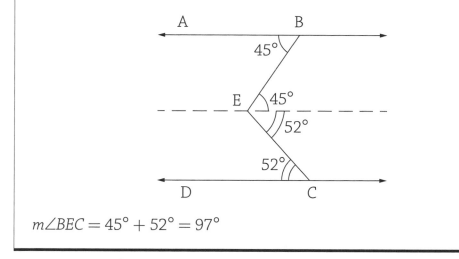

$m\angle BEC = 45° + 52° = 97°$

What if we aren't even given the value of any alternate interior angles? The following example requires you to identify angles that form a linear pair before you look for alternate interior angles.

**EXAMPLE**

▶ In the following diagram, $\overline{AB} \parallel \overline{CD}$, $m\angle MBA = 120°$, and $m\angle PDC = 130°$. Find $m\angle PEM$.

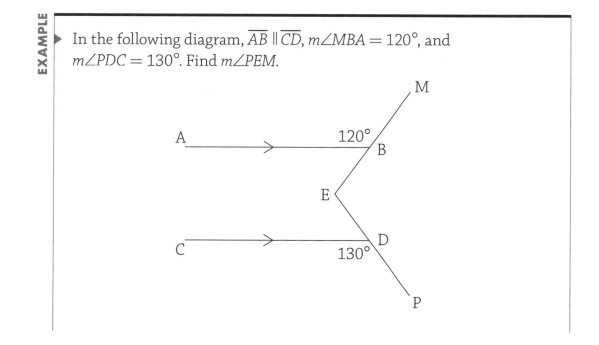

▶ We will begin by drawing an auxiliary line parallel to $\overline{AB}$ through point $E$.

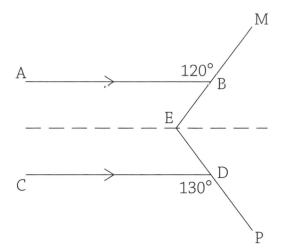

▶ $\angle ABE$ and $\angle ABM$ are supplementary angles because they form a linear pair. This means $m\angle ABE = 60°$. $\angle PDC$ and $\angle EDC$ are supplementary angles because they form a linear pair. This means $m\angle CDE = 50°$.

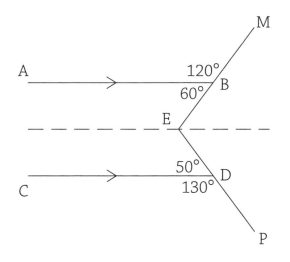

▶ We can now look for congruent alternate interior angles and place their measurements in the diagram.

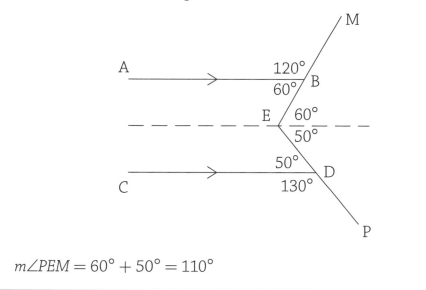

$$m\angle PEM = 60° + 50° = 110°$$

## Proving That the Sum of the Angles of a Triangle Is 180°

One application of an auxiliary line is that we can use it to prove that the sum of the angles of a triangle is 180°. The example that follows shows the statements and reasons used for this proof.

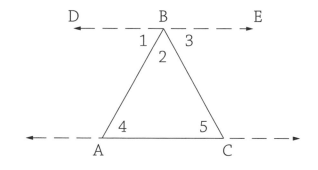

Given: $\triangle ABC$ is drawn with auxiliary line $\overleftrightarrow{DE}$.
$\overline{AC} \parallel \overline{DE}$
Prove: $\angle 2 + \angle 4 + \angle 5 = 180°$

| Satement | Reason |
|---|---|
| 1. $\triangle ABC$ is drawn with auxiliary line $\overleftrightarrow{DE}$, $\overline{AC} \parallel \overline{DE}$. | 1. Given |
| 2. $\angle 1$ and $\angle 4$, $\angle 3$ and $\angle 5$ are alternate interior angles. | 2. Parallel lines cut by a transversal form alternate interior angles. |
| 3. $\angle 1 \cong \angle 4$<br>$\angle 3 \cong \angle 5$ | 3. Alternate interior angles are congruent. |
| 4. $\angle 2 + \angle 1 + \angle 3 = 180°$ | 4. The sum of adjacent angles that form a line is 180°. |
| 5. $\angle 2 + \angle 4 + \angle 5 = 180°$ | 5. Substitution postulate |

In conclusion, we just proved that the three interior angles of a triangle sum to 180°.

## Determining If Lines Are Parallel

We have already learned that parallel lines cut by a transversal form alternate interior angles and corresponding angles that are congruent. In the next part of the chapter, we will discuss the converse of this statement.

*A converse occurs when the hypothesis and conclusion of a conditional statement are switched (flipping a sentence). The converse of "If I practice, then I win" would be "If I win, then I practice."*

**Converse of the Parallel Lines Theorem**

- If two lines are cut by a transversal such that their alternate interior angles are congruent, then the lines are parallel.

- If two lines are cut by a transversal such that their corresponding angles are congruent, then the lines are parallel.

We normally start an example given that two lines are parallel. In the following example, we need to show that the lines are parallel using the given angle measures.

▶ In the following diagram, lines $\ell$ and $m$ are drawn, $m\angle 1 = 5x + 40$, and $m\angle 2 = 10x + 5$. If the value of $x$ is 8, are the lines parallel? What does the value of $x$ have to be so that $\ell \| m$?

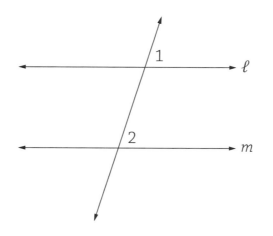

▶ If lines $\ell$ and $m$ are parallel, then $\angle 1$ and $\angle 2$ would be corresponding angles, which means they would be equal. Let's substitute $x = 8$ into both angles and see what they equal.

$$m\angle 1 = 5x + 40 = 5(8) + 40 = 80°$$

$$m\angle 2 = 10x + 5 = 10(8) + 5 = 85°$$

▶ The angles are not equal when $x = 8$; therefore, the lines are *not* parallel when $x = 8$.

▶ To find the value of $x$ that makes the lines parallel, we need to set up an equation that shows that the corresponding angles would be equal.

$$m\angle 1 = m\angle 2$$
$$5x + 40 = 10x + 5$$
$$\underline{-5x - 5 \quad -5x \quad 5}$$
$$\frac{35}{5} = \frac{5x}{5}$$
$$7 = x$$

▶ When $x = 7$, $m\angle 1 = m\angle 2$; therefore, $\ell \| m$.

We hope you now have a good understanding of the converse statement. The following example is similar but a bit more challenging.

▶ In the following diagram, $m\angle JDT = 70°$, $m\angle DMN = 140°$, and $\overline{DM} \cong \overline{RM}$. Is $\overleftrightarrow{JO}$ parallel to $\overleftrightarrow{NA}$?

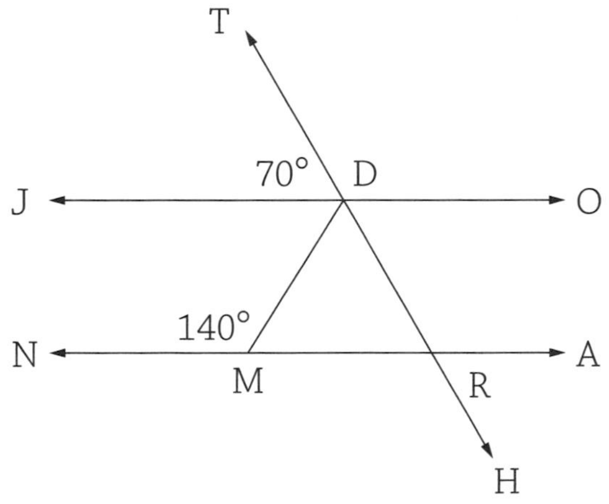

▶ $\angle DMN$ and $\angle DMR$ are supplementary because they form a linear pair. Therefore, $m\angle DMR = 40°$.

▶ Triangle DMR is an isosceles triangle because $\overline{DM} \cong \overline{RM}$. This means that the base angles, $\angle MDR$ and $\angle MRD$, are equal. Because the sum of the angles of a triangle is 180° and the vertex angle is 40°, the base angles equal $\dfrac{180 - 40}{2} = 70°$.

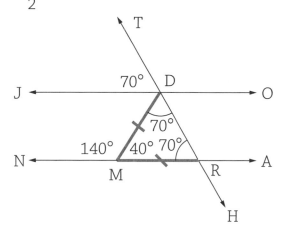

▶ $\overleftrightarrow{JO}$ *is* parallel to $\overleftrightarrow{NA}$ because the corresponding angles $\angle JDT$ and $\angle NRT$ are equal. If two lines are cut by a transversal such that their corresponding angles are congruent, then the lines are parallel. The following diagram shows this clearly by highlighting the "F" that we look for with corresponding angles.

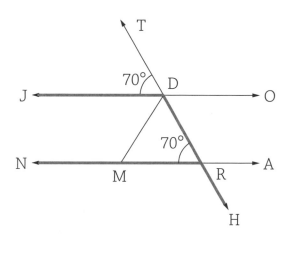

EXAMPLE

The following example shows how to prove line segments parallel in a formal geometric proof.

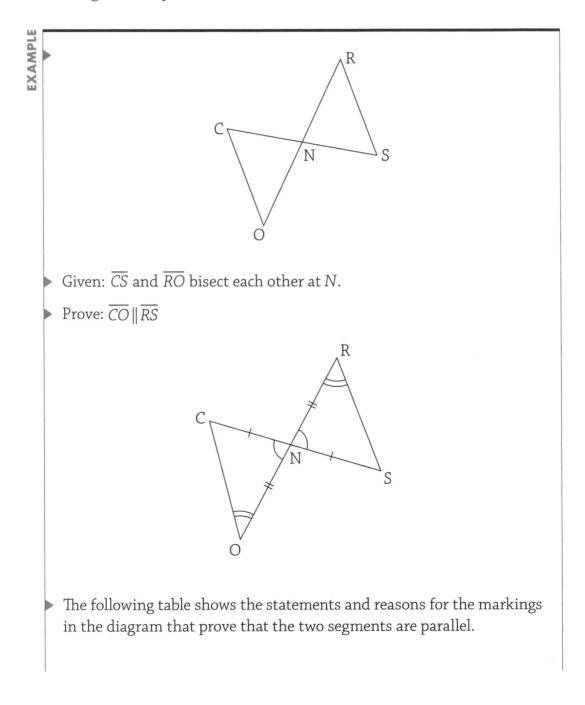

Given: $\overline{CS}$ and $\overline{RO}$ bisect each other at $N$.

Prove: $\overline{CO} \parallel \overline{RS}$

The following table shows the statements and reasons for the markings in the diagram that prove that the two segments are parallel.

| Statement | Reason |
|---|---|
| 1. $\overline{CS}$ and $\overline{RO}$ bisect each other at $N$. | 1. Given |
| 2. $\overline{CN} \cong \overline{SN}$ <br> $\overline{RN} \cong \overline{ON}$ | 2. A bisector divides a segment into two congruent segments. |
| 3. $\angle CNO$ and $\angle SNR$ are vertical angles. | 3. Intersecting lines form vertical angles. |
| 4. $\angle CNO \cong \angle SNR$ | 4. Vertical angles are congruent. |
| 5. $\triangle CNO \cong \triangle SNR$ | 5. SAS $\cong$ SAS |
| 6. $\angle CON \cong \angle SRN$ | 6. If two triangles are congruent, then their corresponding parts are congruent. |
| 7. $\overline{CO} \parallel \overline{RS}$ | 7. If two lines are cut by a transversal such that their alternate interior angles are congruent, then the lines are parallel. |

The proof just completed showed that the lines were parallel because the alternate interior angles were congruent. The following are four methods that can be used to prove lines parallel. You will have the opportunity to practice more of these in the practice section at the conclusion of the chapter.

Two lines are parallel if they are cut by a transversal such that their:

- alternate interior angles are congruent,
- corresponding angles are congruent,
- alternate exterior angles are congruent, or
- consecutive interior angles are supplementary.

Two lines are also parallel if they are perpendicular to the same line and lie on the same plane.

## EXERCISES

### EXERCISE 7-1

*Use the following diagram, of two parallel lines cut by a transversal, to determine if the pairs of angles in the following questions are congruent.*

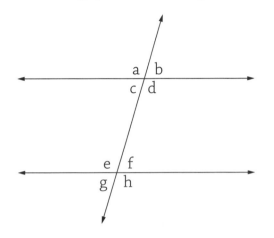

1. $\angle c$ and $\angle f$

2. $\angle b$ and $\angle g$

3. $\angle f$ and $\angle h$

4. $\angle d$ and $\angle h$

5. $\angle a$ and $\angle g$

6. $\angle c$ and $\angle e$

7. $\angle e$ and $\angle h$

## EXERCISE 7-2

*Use the following diagram of* $\overline{AB} \parallel \overline{CD}$ *to answer the questions in this exercise.*

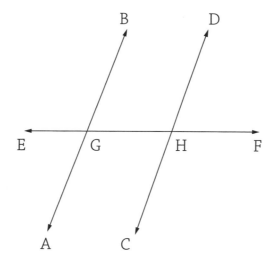

1. If $m\angle AGH = 108°$, what is $m\angle DHG$?

2. If $m\angle CHG = 3x + 2$ and $m\angle AGH = 5x + 18$, what is the value of $x$?

3. If $m\angle BGH = 3x + 20$ and $m\angle CHG = x + 32$, what is $m\angle BGH$?

4. If $m\angle EGB = 6x - 14$ and $m\angle GHD = 3x + 61$, what is $m\angle GHD$?

5. If $m\angle AGH = 4x - 45$ and $m\angle DHF = 2x - 15$, what is the value of $x$?

## EXERCISE 7-3

*Use the following diagram to answer questions 1–4.*

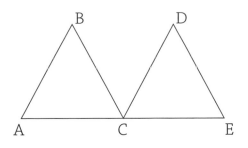

1. $m\angle BAC = 4x - 45$ and $m\angle DCE = 2x + 15$. If $x$ equals 20, is $\overline{AB} \parallel \overline{CD}$?

2. $m\angle DEC = 10x - 10$ and $m\angle BCA = 2x + 14$. If $x$ equals 3, is $\overline{BC} \parallel \overline{DE}$?

3. If $\overline{AB} \parallel \overline{CD}$, $m\angle A = x + 27$, and $m\angle DCE = 2x - 18$, what is $m\angle A$?

4. Given: $\overline{AB} \parallel \overline{CD}$, $\overline{DE} \parallel \overline{BC}$, and $C$ is the midpoint of $\overline{AE}$. Prove: $\angle B \cong \angle D$.

## EXERCISE 7-4

*Use the following diagram of $\triangle MAT$, with $\overline{MA} \parallel \overline{TH}$, $m\angle M = 101°$, and $m\angle ATH = 43°$, to answer the following two questions.*

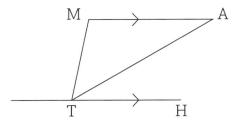

1. Find $m\angle MAT$.

2. Find $m\angle ATM$.

## EXERCISE 7-5

*This exercise requires you to complete the missing steps in the following proof.*

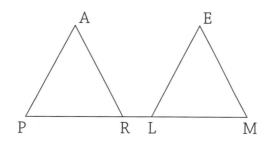

Given: $\overline{PA} \cong \overline{LE}$
$\overline{RA} \cong \overline{ME}$
$\overline{PL} \cong \overline{MR}$

Prove: $\overline{PA} \parallel \overline{LE}$

| Statement | Reason |
|---|---|
| $\overline{PA} \cong \overline{LE}$, $\overline{RA} \cong \overline{ME}$ | Given |
| $\overline{PL} \cong \overline{MR}$ | Given |
| 1. | Reflexive property |
| $\overline{PR} \cong \overline{ML}$ | 2. |
| $\triangle APR \cong \triangle ELM$ | 3. |
| $\angle APR \cong \angle ELM$ | If two triangles are congruent, then their corresponding angles are congruent. |
| $\overline{PA} \parallel \overline{LE}$ | 4. |

## EXERCISE 7-6

*Use the following diagrams to answer these questions.*

1. $\overline{GM} \parallel \overline{SA}$. $m\angle MGI = 35°$ and $m\angle ASI = 61°$. Find $m\angle SIG$.

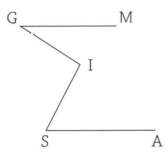

2. $\overline{PN} \parallel \overline{ST}$. $m\angle RPN = 128°$ and $m\angle RST = 120°$. Find $m\angle PRS$.

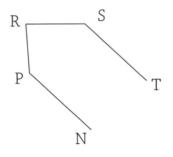

Flashcard App

 **Parallelograms**

## MUST KNOW

 Quadrilaterals are four-sided polygons.

 A parallelogram is a special kind of quadrilateral in which both pairs of opposite sides are parallel.

A rectangle is a parallelogram in which all angles are right and the diagonals are congruent.

A rhombus is a parallelogram in which all four sides are congruent. Its diagonals are perpendicular to each other and bisect the angles.

A square is a parallelogram that displays all the properties of a rectangle and a rhombus.

**A** **quadrilateral** is a four-sided polygon. Everywhere you look you can find quadrilaterals. Our cell phones and laptops, football fields, and basketball courts all take on the quadrilateral shape. This chapter will be focusing on **parallelograms**. Parallelograms are quadrilaterals that contain the following special properties:

- Opposite sides are parallel.

- Opposite sides are congruent.

- Opposite angles are congruent.

- Consecutive angles are supplementary.

- Diagonals bisect each other.

- The diagonal divides the parallelogram into two congruent triangles:

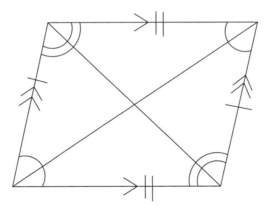

Let's look at an example in which we solve for an unknown angle by taking advantage of the fact that opposite sides of a parallelogram are parallel.

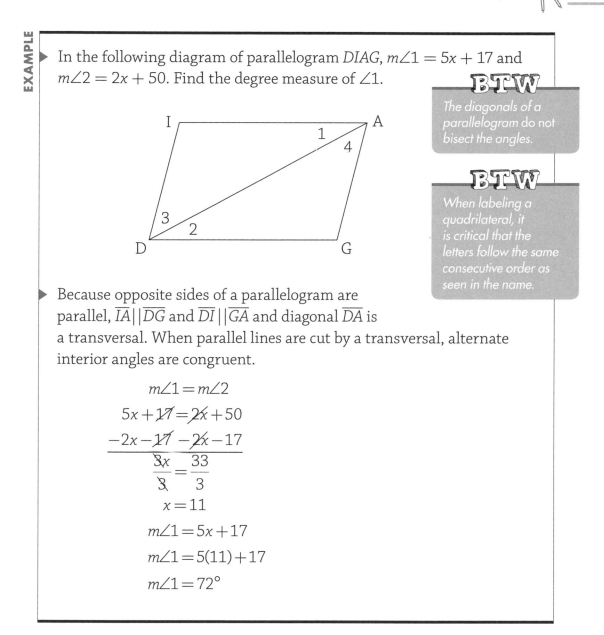

In the following diagram of parallelogram $DIAG$, $m\angle 1 = 5x + 17$ and $m\angle 2 = 2x + 50$. Find the degree measure of $\angle 1$.

> **BTW**
> The diagonals of a parallelogram do not bisect the angles.

> **BTW**
> When labeling a quadrilateral, it is critical that the letters follow the same consecutive order as seen in the name.

Because opposite sides of a parallelogram are parallel, $\overline{IA} \parallel \overline{DG}$ and $\overline{DI} \parallel \overline{GA}$ and diagonal $\overline{DA}$ is a transversal. When parallel lines are cut by a transversal, alternate interior angles are congruent.

$$m\angle 1 = m\angle 2$$
$$5x + 17 = 2x + 50$$
$$\underline{-2x - 17 \quad -2x - 17}$$
$$\frac{3x}{3} = \frac{33}{3}$$
$$x = 11$$
$$m\angle 1 = 5x + 17$$
$$m\angle 1 = 5(11) + 17$$
$$m\angle 1 = 72°$$

The opposite angles of a parallelogram are congruent. Let's use this property to solve for an angle of the parallelogram in the following example.

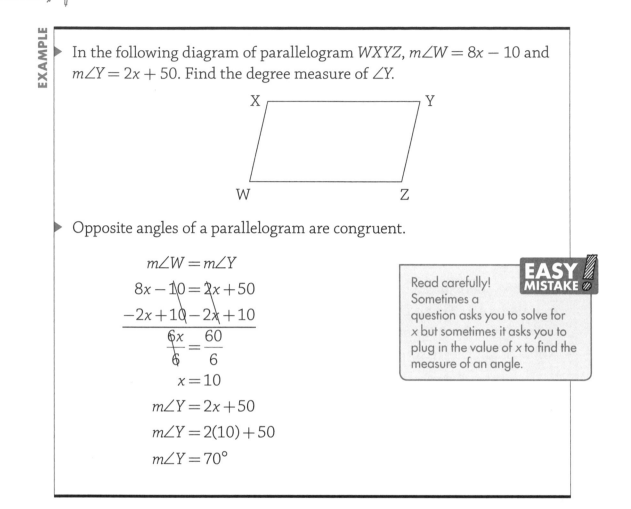

EXAMPLE

In the following diagram of parallelogram $WXYZ$, $m\angle W = 8x - 10$ and $m\angle Y = 2x + 50$. Find the degree measure of $\angle Y$.

Opposite angles of a parallelogram are congruent.

$$m\angle W = m\angle Y$$
$$8x - 10 = 2x + 50$$
$$-2x + 10 \quad -2x + 10$$
$$\frac{6x}{6} = \frac{60}{6}$$
$$x = 10$$
$$m\angle Y = 2x + 50$$
$$m\angle Y = 2(10) + 50$$
$$m\angle Y = 70°$$

**EASY MISTAKE**

Read carefully! Sometimes a question asks you to solve for $x$ but sometimes it asks you to plug in the value of $x$ to find the measure of an angle.

In the next example, we'll solve for the length of a segment using the fact that opposite sides of a parallelogram are congruent.

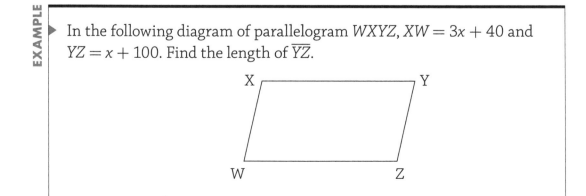

EXAMPLE

In the following diagram of parallelogram $WXYZ$, $XW = 3x + 40$ and $YZ = x + 100$. Find the length of $\overline{YZ}$.

▸ The opposite sides of a parallelogram are congruent.

$$XW = YZ$$
$$3x + 40 = x + 100$$
$$\underline{-x - 40 \quad -x - 40}$$
$$\frac{2x}{2} = \frac{60}{2}$$
$$x = 30$$
$$YZ = x + 100$$
$$YZ = 30 + 100$$
$$YZ = 130$$

Now let's solve for a variable using the fact that the consecutive angles of a parallelogram are supplementary.

▸ In the following diagram of parallelogram $WXYZ$, $m\angle X = 4x - 10$ and $m\angle W = x + 50$. Find the value of $x$.

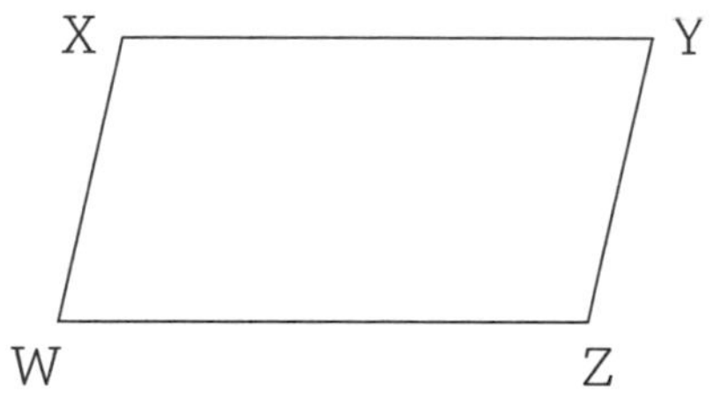

▸ Consecutive angles of a parallelogram are supplementary.

$$m\angle X + m\angle W = 180°$$
$$4x - 10 + x + 50 = 180$$
$$5x + 40 = 180$$
$$\underline{-40 \quad -40}$$
$$\frac{5x}{5} = \frac{140}{5}$$
$$x = 28$$

We are going to look at an example that involves the fact that the diagonals of a parallelogram bisect each other. But first we should look at a geometric proof of the property.

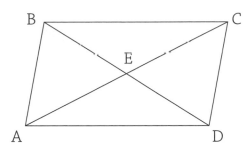

Given: $ABCD$ is a parallelogram.
Prove: $\overline{BD}$ and $\overline{CA}$ bisect each other at $E$.

| Statement | Reason |
|---|---|
| 1. $ABCD$ is a parallelogram. | 1. Given |
| 2. $\overline{AD} \cong \overline{BC}$ | 2. Opposite sides of a parallelogram are congruent. |
| 3. $\overline{AD} \parallel \overline{BC}$ | 3. Opposite sides of a parallelogram are parallel. |
| 4. $\angle CBD$ and $\angle ADB$, $\angle BCA$ and $\angle DAC$ are alternate interior angles. | 4. Parallel lines cut by a transversal form alternate interior angles. |
| 5. $\angle CBD \cong \angle ADB$ $\angle BCA \cong \angle DAC$ | 5. Alternate interior angles are congruent. |
| 6. $\triangle BEC \cong \triangle DEA$ | 6. ASA $\cong$ ASA |
| 7. $\overline{BE} \cong \overline{DE}$, $\overline{AE} \cong \overline{CE}$ | 7. Corresponding parts of congruent triangles are congruent (CPCTC). |
| 8. $E$ is the midpoint of $\overline{BD}$ and $\overline{AC}$. | 8. A midpoint divides a segment into two congruent segments. |
| 9. $\overline{BD}$ and $\overline{CA}$ bisect each other at $E$. | 9. A bisector intersects a segment at its midpoint. |

## EXTRA HELP

Now that we have proved that the diagonals of a parallelogram bisect each other, let's apply this property algebraically in the next example.

**EXAMPLE**

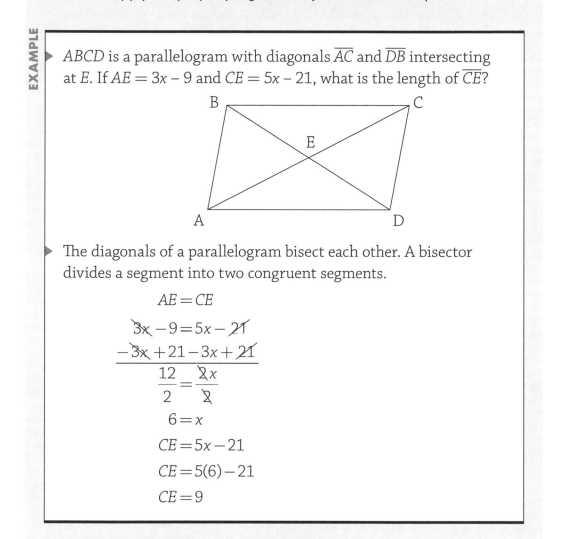

$ABCD$ is a parallelogram with diagonals $\overline{AC}$ and $\overline{DB}$ intersecting at $E$. If $AE = 3x - 9$ and $CE = 5x - 21$, what is the length of $\overline{CE}$?

The diagonals of a parallelogram bisect each other. A bisector divides a segment into two congruent segments.

$$AE = CE$$

$$3x - 9 = 5x - 21$$

$$\underline{-3x + 21 \quad -3x + 21}$$

$$\frac{12}{2} = \frac{2x}{2}$$

$$6 = x$$

$$CE = 5x - 21$$

$$CE = 5(6) - 21$$

$$CE = 9$$

It is time for us to look at some special parallelograms. These parallelograms contain all the basic parallelogram properties but also have extra properties of their own.

# Rectangles

A **rectangle** is a parallelogram that contains a right angle. If you're thinking there are four right angles in a rectangle, then you are correct. The definition says that it is a *parallelogram* with a right angle. We know that opposite angles are congruent and consecutive angles are supplementary in a parallelogram. This means the opposite angle is a right angle and the consecutive angles are also right angles. Therefore, all four angles in a parallelogram are right angles.

The special properties of a rectangle include:

- all the properties of a parallelogram,

- four right angles, and

- congruent diagonals.

Let's take a moment to prove the third property of a rectangle—that the diagonals of a rectangle are congruent.

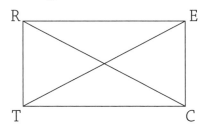

Given: Rectangle *RECT*

Prove: $\overline{RC} \cong \overline{ET}$

| Statement | Reason |
|---|---|
| 1. *RECT* is a rectangle. | 1. Given |
| 2. $\overline{RT} \cong \overline{EC}$ | 2. Opposite sides of a rectangle are congruent. |
| 3. $\overline{TC} \cong \overline{TC}$ | 3. Reflexive property |
| 4. $\angle RTC \cong \angle ECT$ | 4. Angles of a rectangle are congruent right angles. |
| 5. $\triangle RTC \cong \triangle ECT$ | 5. SAS $\cong$ SAS |
| 6. $\overline{RC} \cong \overline{ET}$ | 6. CPCTC |

Now that we have seen a proof showing that the diagonals of a rectangle are congruent, let's apply that property toward an algebraic problem.

**EXAMPLE**

▶ In rectangle $ABCD$, diagonal $AC = 2y + 8$ and diagonal $BD = 4y - 10$. Find the length of $\overline{AC}$.

▶ The diagonals of a rectangle are congruent.

$$AC = BD$$
$$2y + 8 = 4y - 10$$
$$\underline{-2y + 10 \qquad -2y + 10}$$
$$\frac{18}{2} = \frac{2y}{2}$$
$$9 = y$$
$$AC = 2y + 8$$
$$AC = 2(9) + 8$$
$$AC = 26$$

Because the diagonal of a parallelogram divides the parallelogram into two congruent triangles and a rectangle is a parallelogram containing four right angles, the diagonal of a rectangle divides the rectangle into two congruent right triangles. We will use this property in the following example.

**EXAMPLE**

▶ Parallelogram $RECT$ has sides that are 7 feet and 24 feet long. The diagonal is 25 feet. Is $RECT$ a rectangle?

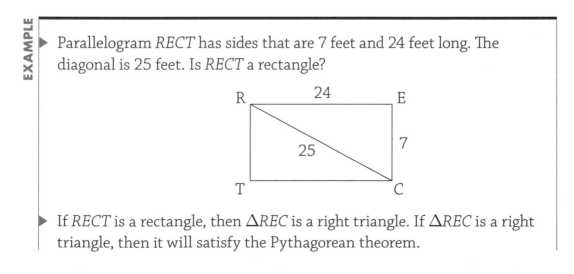

▶ If $RECT$ is a rectangle, then $\triangle REC$ is a right triangle. If $\triangle REC$ is a right triangle, then it will satisfy the Pythagorean theorem.

$$a^2 + b^2 = c^2$$
$$(7)^2 + (24)^2 \overset{?}{=} (25)^2$$
$$625 = 625$$

▶ The lengths of the sides of the triangle satisfy the Pythagorean theorem, which means $\angle E$ is a right angle. If a parallelogram contains a right angle, then it is a rectangle; therefore, $RECT$ is a rectangle.

## Rhombuses

A **rhombus** is also a parallelogram that contains extra properties. The extra properties of a rhombus are different from the extra properties of a rectangle. These properties are:

- A rhombus includes all the properties of a parallelogram.

- All four sides are congruent.

- Diagonals are perpendicular to each other.

- Diagonals bisect the angles.

> **BTW**
>
> If the diagonals of a rhombus are perpendicular to each other, then they form right angles at their point of intersection.

It is time to apply the properties of a rhombus to algebraic problems. Our first example involves the fact that the diagonals of a rhombus bisect each other.

**EXAMPLE**

▶ The following diagram of $RHOM$ represents a rhombus with diagonals $\overline{HM}$ and $\overline{RO}$ intersecting at B. If $m\angle HRO = x + 42$, and $m\angle MRO = 2x + 12$, what is $m\angle MRO$?

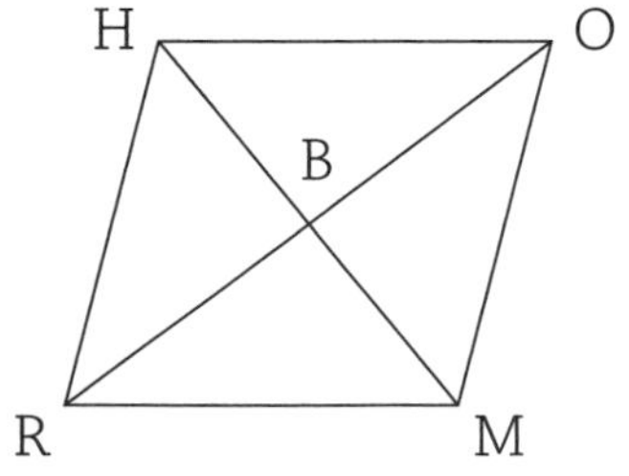

▶ The diagonals of a rhombus bisect the angles. A bisector divides an angle into two congruent angles.

$$m\angle HRO = m\angle MRO$$
$$x + 42 = 2x + \cancel{12}$$
$$\underline{-x - 12 \quad -x - \cancel{12}}$$
$$30 = x$$
$$m\angle MRO = 2x + 12$$
$$m\angle MRO = 2(30) + 12$$
$$m\angle MRO = 72°$$

The next example involves many properties of a rombus: the diagonals of a rhombus are perpendicular, the diagonals of a rhombus bisect each other, and all sides of a rhombus are congruent.

**EXAMPLE**

▶ The following diagram of *RHOM* represents a rhombus with diagonals $\overline{HM}$ and $\overline{RO}$ intersecting at *B*. If $HM = 10$ and $RO = 24$, what is the perimeter of rhombus *RHOM*?

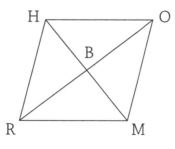

▶ The diagonals of a rhombus bisect each other, which is why $HB = BM = 5$ and $RB = BO = 12$.

▶ The diagonals of a rhombus are perpendicular, which is why the rhombus is divided into four right triangles. The hypotenuse of the right triangle is the side of the rhombus.

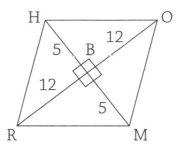

▶ To find the perimeter of the rhombus, we need to know the length of one of the sides of the rhombus. We can use the Pythagorean theorem.

$$a^2 + b^2 = c^2$$
$$(5)^2 + (12)^2 = (HO)^2$$
$$25 + 144 = (HO)^2$$
$$\sqrt{169} = \sqrt{(HO)^2}$$
$$13 = HO$$

▶ All sides of a rhombus are congruent, which means the lengths of each side of *RHOM* is 13.

▶ The perimeter of a rhombus is the sum of all the sides of the rhombus. Therefore, the perimeter of *RHOM* is $4(13) = 52$.

## Squares

The **square** is the "most special" parallelogram because it contains all the properties of a rectangle and a rhombus. This means that a square is a rectangle and also a rhombus. But the reverse is not the case. A rhombus is *not* always a square and a rectangle is *not* always a square. The following diagram shows the relationships of different quadrilaterals.

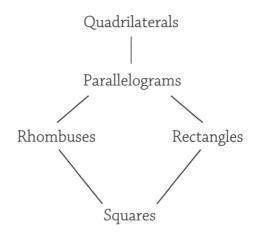

### Properties of a Square

- All the properties of a parallelogram

- All the properties of a rectangle

- All the properties of a rhombus

Let's take a moment to apply some of the properties of a square. Our first example involves the fact that the diagonals of a square are congruent.

**EXAMPLE**

▶ In square *JONA*, $JN = c + 25$ and $OA = 15c - 3$. Find the value of *c*.

▶ Let's begin by drawing the diagram.

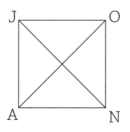

> The diagonals of a square are congruent.

$$JN = OA$$

$$\not{c} + 25 = 15c - \not{3}$$

$$\underline{-\not{c} + 3 \quad -c + \not{3}}$$

$$\frac{28}{14} = \frac{\cancel{14}c}{\cancel{14}}$$

$$2 = c$$

## EXTRA HELP

We learned that the sides of a square are congruent and that a square contains four right angles. Let's use those facts to find the length of the diagonal of the square in the following example.

> In square $BASE$, $BA = x + 4$ and $AS = 2x + 2$. Find the length of diagonal $\overline{SB}$.

> Let's begin by drawing a diagram of square $BASE$. Because a square contains four right angles, we will include that in our diagram.

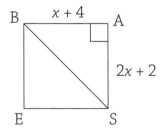

▶ All sides of a square are congruent. We can set the expressions for the sides equal, which will give us the ability to find the length of the sides of the square.

$$BA = AS$$
$$\cancel{x} + 4 = 2x + \cancel{2}$$
$$\underline{-\cancel{x} - 2 \quad -x - \cancel{2}}$$
$$2 = x$$
$$BA = x + 4$$
$$BA = 2 + 4$$
$$BA = 6$$

▶ Now that we know all the sides of the square are 6, we can use the Pythagorean theorem to solve for the diagonal of the square.

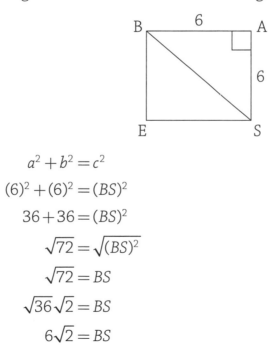

$$a^2 + b^2 = c^2$$
$$(6)^2 + (6)^2 = (BS)^2$$
$$36 + 36 = (BS)^2$$
$$\sqrt{72} = \sqrt{(BS)^2}$$
$$\sqrt{72} = BS$$
$$\sqrt{36}\sqrt{2} = BS$$
$$6\sqrt{2} = BS$$

## Trapezoids

Last but not least, it is time to discuss the properties of a **trapezoid**. Some mathematicians define a trapezoid as a quadrilateral with *only one* pair of parallel sides, and others define it as a quadrilateral with *at least* one pair of parallel sides. So, is a parallelogram a trapezoid? Some argue that it is a trapezoid because a parallelogram has at least one pair of parallel sides. Those who define a trapezoid as having only one pair of parallel sides believe it is not a parallelogram. Either way, the following diagrams are examples of trapezoids.

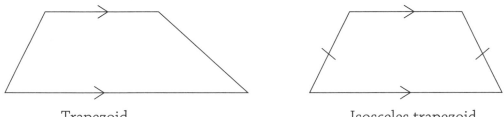

Trapezoid                    Isosceles trapezoid

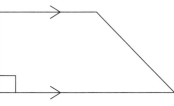

Right trapezoid

An **isosceles trapezoid** contains the following properties:

- nonparallel sides are congruent,
- base angles are congruent, and
- diagonals are congruent.

A **right trapezoid** is a trapezoid that has two adjacent right angles.

Let's apply the properties of an isosceles trapezoid towards algebraic problems. Our first example involves the fact that the diagonals of an isosceles trapezoid are congruent.

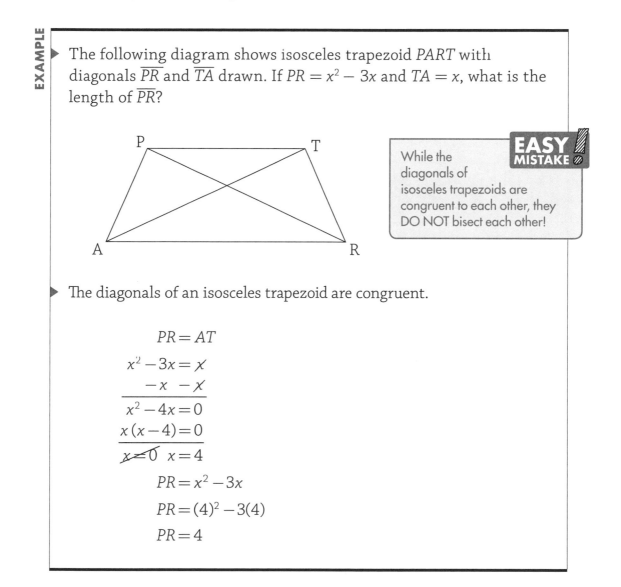

**EXAMPLE**

▶ The following diagram shows isosceles trapezoid $PART$ with diagonals $\overline{PR}$ and $\overline{TA}$ drawn. If $PR = x^2 - 3x$ and $TA = x$, what is the length of $\overline{PR}$?

**EASY MISTAKE**

While the diagonals of isosceles trapezoids are congruent to each other, they DO NOT bisect each other!

▶ The diagonals of an isosceles trapezoid are congruent.

$$PR = AT$$
$$x^2 - 3x = x$$
$$\underline{-x \quad -x}$$
$$x^2 - 4x = 0$$
$$x(x - 4) = 0$$
$$x = 0 \quad x = 4$$
$$PR = x^2 - 3x$$
$$PR = (4)^2 - 3(4)$$
$$PR = 4$$

The next example looks at the fact that the bases of a trapezoid are parallel to solve for an angle.

▶ The following diagram shows isosceles trapezoid *PART* with diagonals $\overline{PR}$ and $\overline{TA}$ drawn. If $m\angle PRA = x + 52$ and $m\angle RPT = 2x + 18$, what is $m\angle RPT$?

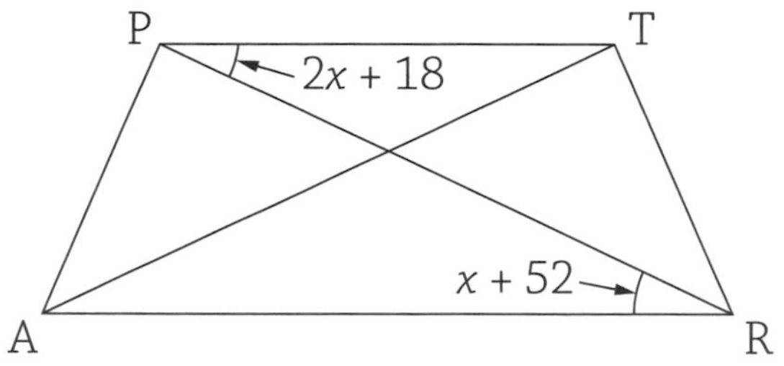

▶ The bases of a trapezoid are parallel, which means $\overline{PT}||\overline{AR}$. When parallel lines are cut by a transversal, they form congruent alternate interior angles.

$$m\angle PRA = m\angle RPT$$
$$\cancel{x} + 52 = 2x + \cancel{18}$$
$$\underline{-\cancel{x} - 18 \quad -x - \cancel{18}}$$
$$34 = x$$
$$m\angle RPT = 2x + 18$$
$$m\angle RPT = 2(34) + 18$$
$$m\angle RPT = 86°$$

This example involves the fact that the base angles of an isosceles trapezoid are congruent to solve for a variable.

EXAMPLE

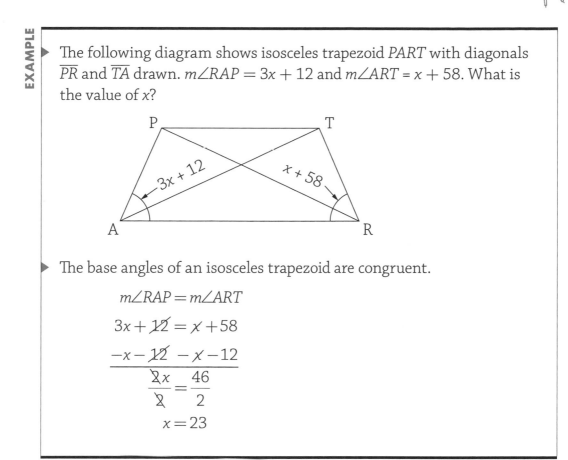

▶ The following diagram shows isosceles trapezoid *PART* with diagonals $\overline{PR}$ and $\overline{TA}$ drawn. $m\angle RAP = 3x + 12$ and $m\angle ART = x + 58$. What is the value of $x$?

▶ The base angles of an isosceles trapezoid are congruent.

$$m\angle RAP = m\angle ART$$

$$3x + \cancel{12} = x + 58$$

$$\underline{-x - \cancel{12} \quad -x - 12}$$

$$\frac{\cancel{2}x}{\cancel{2}} = \frac{46}{2}$$

$$x = 23$$

## Median of a Trapezoid

The **median of a trapezoid** connects the midpoints of the nonparallel sides of the trapezoid. The median is parallel to both bases of the trapezoid. The length of the median of a trapezoid is the average of the length of both bases of the trapezoid.

Let's practice using the median of a trapezoid to solve algebraic problems. Our first example uses the lengths of the parallel bases to find the length of the median of the trapezoid.

> **BTW**
>
> *The median of a trapezoid is also sometimes called the **midsegment** of the trapezoid.*

▶ Trapezoid $ABCD$ is drawn. $M$ is the midpoint of $\overline{AB}$ and $N$ is the midpoint of $\overline{CD}$. $BC = 23$ and $AD = 34$. Find the length of $\overline{MN}$.

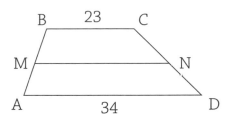

▶ The median of a trapezoid is equal to the average of the lengths of the parallel bases of the trapezoid.

$$MN = \frac{BC + AD}{2}$$

$$MN = \frac{23 + 34}{2}$$

$$MN = \frac{57}{2}$$

$$MN = 28.5$$

Our next example is a more challenging algebraic problem. It allows us to take advantage of the properties of the median of a trapezoid to solve for a variable.

▶ Trapezoid $ABCD$ is drawn. $M$ is the midpoint of $\overline{AB}$ and $N$ is the midpoint of $\overline{CD}$. $AD = 3x + 12$, $BC = x$, and $MN = 20$. Find the value of $x$.

▶ The median of a trapezoid is equal to the average of the lengths of the parallel bases.

$$MN = \frac{BC + AD}{2}$$

$$20 = \frac{x + 3x + 12}{2}$$

$$20 = 4x + 12$$

$$40 = 4x + 12$$

$$\begin{array}{cc} -12 & -12 \end{array}$$

$$\frac{28}{4} = \frac{4x}{4}$$

$$7 = x$$

## EXERCISES

### EXERCISE 8-1

*Use the information given to answer these two questions.*

1.  In parallelogram $ABCD$, $BC = 22x - 1$ and $AD = 3x + 37$. Find the length of $\overline{AD}$.

2.  In parallelogram $ABCD$, $m\angle A = 37°$. Find $m\angle B$ and $m\angle C$.

### EXERCISE 8-2

*Determine whether each of the following statements is true or false.*

1.  The diagonals of all parallelograms bisect the angles.

2.  The diagonals of a rhombus are congruent.

3.  The diagonals of a square are congruent and perpendicular.

4.  The diagonal divides a parallelogram into two congruent triangles.

5.  Consecutive angles of a parallelogram are supplementary.

6.  A rectangle is a square.

7.  A square is a rectangle.

8.  The diagonals of an isosceles trapezoid are congruent.

9.  The diagonals of all parallelograms bisect each other.

10. A square has four congruent sides and four congruent right angles.

## EXERCISE 8-3

*Use the following diagram and given information to solve this group of questions. In rhombus* RSTW, *the diagonals intersect at M.*

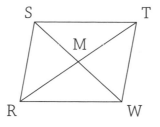

1.  If $RM = 12$ and $SM = 9$, what is the length of $\overline{RS}$?

2.  If $RT = 24$ and $SW = 10$, what is the perimeter of rhombus $RSTW$?

3.  The perimeter of the rhombus is 100, and $WT = 2x + 1$. Find the value of $x$.

4.  $m\angle SMT = 3x - 15$. Find the value of $x$.

5.  $m\angle MRS = 40°$. Find $m\angle MSR$.

## EXERCISE 8-4

*Use the following diagram and given information to solve the questions in this exercise. In rectangle* ABCD, *diagonals* $\overline{AC}$ *and* $\overline{BD}$ *are drawn.*

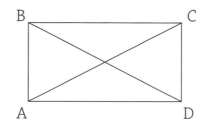

1.  $AC = x^2 - 2x - 11$ and $BD = 2x + 1$. Find the length of $\overline{BD}$.

2.  $m\angle DAB = 7x - 8$. Find the value of $x$.

3.  $AB = 15$ and $AC = 39$. Find the perimeter of rectangle $ABCD$.

## EXERCISE 8-5

*Use the following diagram and the related information to solve these three questions. In isosceles trapezoid GLAD, diagonals $\overline{GA}$ and $\overline{DL}$ are drawn.*

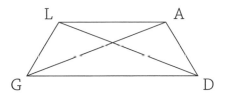

1.  $LD = 17b + 3$ and $AG = 88$. Find the value of $b$.

2.  $m\angle LGD = 40°$. Find $m\angle GLA$.

3.  $m\angle ALD = y + 30$ and $m\angle GDL = 110°$. Find the value of $y$.

## EXERCISE 8-6

*Use the following diagram and the given information to solve the questions in this exercise. Trapezoid MDIN with median $\overline{EA}$ is drawn.*

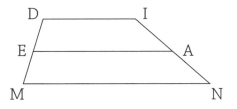

1.  $DI = 17$ and $MN = 25$. Find the length of $\overline{EA}$.

2.  $DI = 33$ and $EA = 59$. Find the length of $\overline{MN}$.

3.  $DI = x + 4$, $MN = 3x + 6$, and $EA = x + 7$. Find the value of $x$.

Flashcard
App

# 9 Coordinate Geometry

## MUST KNOW

 Coordinate geometry allows us to describe geometric shapes algebraically by assigning mathematical values to points on—commonly—an *x–y* axis.

 The distance formula, which is derived from the Pythagorean theorem, can be used to show that segments are the same length.

The slope formula can determine if lines are parallel or perpendicular. Parallel lines have equal slopes, whereas perpendicular lines have slopes that are negative reciprocals.

 The midpoint formula can be used to prove that the diagonals of a parallelogram bisect each other.

The formulas for distance, midpoint, and slope can be used to prove the properties of lines and geometric figures.

oordinate geometry is the link between geometry and algebra. It uses algebra to study geometric properties. Coordinate geometry provides us with the ability to write the equations of lines and algebraically show the properties of all the triangles and quadrilaterals we learned about in previous chapters.

Coordinates give the location of points on the Cartesian plane using the x- and y-axes.

**IRL**   The invention of Cartesian coordinates in the seventeenth century by **René Descartes** revolutionized mathematics by linking geometry and algebra.

Let's begin by reviewing how to plot a point. Every point is made up of an $x$ coordinate (also called the **abscissa**) and a $y$ coordinate (also called the **ordinate**). The $x$ coordinate tells the horizontal location of the point, and the $y$ coordinate tells the vertical location of the point.

**EXAMPLE**

▶ The following diagram shows three plotted points: $A(2,1)$, $B(5,1)$, and $C(2,7)$.

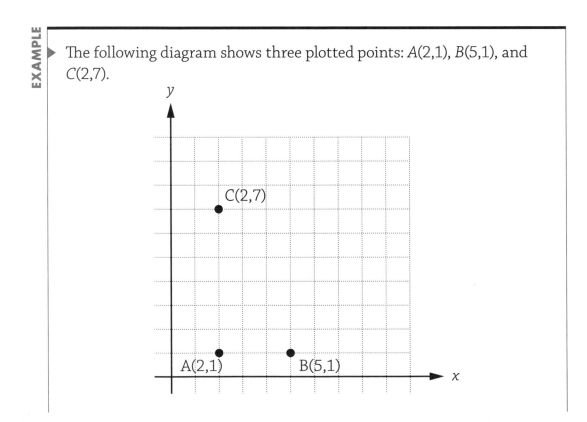

▶ Can you find the exact distance between these points? It is easy to do when the two points form a horizontal or vertical line. For example, when you plot point $A(2,1)$ and point $B(5,1)$, you can actually count the boxes between them and see a horizontal distance of three units between them. You can also see that $AC = 6$, meaning there is a vertical distance of six units between points $A$ and $C$.

▶ How can we find the length of $\overline{BC}$ if the points do not form a vertical or a horizontal line? We can draw a right triangle where $\overline{BC}$ is the hypotenuse and use the Pythagorean theorem to find the length of the hypotenuse with the legs of the right triangle consisting of the vertical segment and horizontal segment.

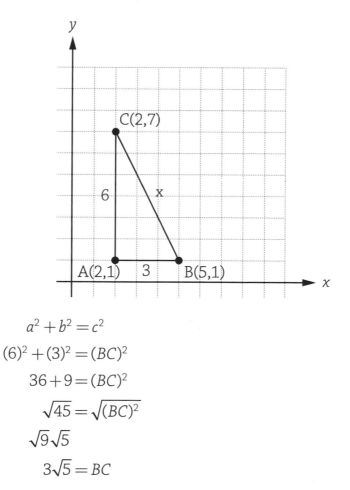

$$a^2 + b^2 = c^2$$
$$(6)^2 + (3)^2 = (BC)^2$$
$$36 + 9 = (BC)^2$$
$$\sqrt{45} = \sqrt{(BC)^2}$$
$$\sqrt{9}\sqrt{5}$$
$$3\sqrt{5} = BC$$

## Distance Formula

The distance formula is used to find the length between two points on a Cartesian plane. It is derived from using the Pythagorean theorem to find the distance between two points $A(x_1,y_1)$ and $C(x_2,y_2)$.

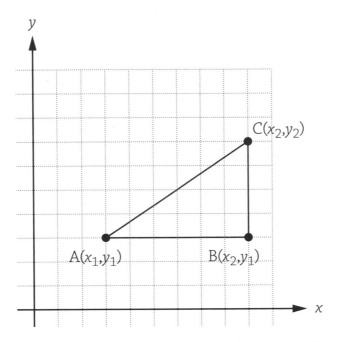

Just as we did in the previous example, we will draw a right triangle such that $\overline{AC}$ represents the hypotenuse and then use the Pythagorean theorem to find the length of $\overline{AC}$.

$$(AB)^2 + (BC)^2 = (AC)^2$$

- The length of $AB$ is represented by $(x_2 - x_1)$.

- The length of $BC$ is represented by $(y_2 - y_1)$.

We substitute these lengths into the Pythagorean theorem and get:

$$(x_2 - x_1)^2 + (y_2 - y_1)^2 = (AC)^2$$

To find the length of *AC*, we square root both sides:

$$\sqrt{(x_2 - x_1)^2 + (y_2 - y_1)^2} = AC$$

Therefore, to find the distance between any two points, we can use the **distance formula**, which is $d = \sqrt{(x_2 - x_1)^2 + (y_2 - y_1)^2}$.

Let's practice finding the distance between two points using the distance formula in the following example.

**EXAMPLE**

▶ $\overline{RS}$ has endpoints $R(-4,5)$ and $S(8,2)$. Use the distance formula to find the length of $\overline{RS}$.

▶ Let $(x_1, y_1)$ be $(-4,5)$ and $(x_2, y_2)$ be $(8,2)$

$$
\begin{aligned}
RS &= \sqrt{(x_2 - x_1)^2 + (y_2 - y_1)^2} \\
&= \sqrt{(8 - (-4))^2 + (2 - 5)^2} \\
&= \sqrt{(12)^2 + (-3)^2} \\
&= \sqrt{144 + 9} \\
&= \sqrt{153}
\end{aligned}
$$

## Using the Distance Formula to Classify Shapes

Now that we know how to use the distance formula, let's use it to show the special properties of the different triangles and parallelograms we learned about in previous chapters of this book. In Chapter 3, we learned how to classify triangles as isosceles, equilateral, and right using their side lengths. In Chapter 8, we learned how to classify quadrilaterals as parallelograms, squares, rhombuses, and rectangles using their side length. Because the distance formula finds the length between two points, we must be able to use it to classify these different figures if we are given the coordinates for the vertices of the figure. Let's try this in the next example.

▶ Consider $\triangle MAT$ with $M(1,1)$, $A(4,6)$, and $T(7,11)$. Using coordinate geometry, prove that $\triangle MAT$ is an isosceles triangle.

▶ One property of an isosceles triangle is that two of the sides are congruent. We can use the distance formula to find the lengths of the sides of the triangle and show that two sides have the same length. For visual purposes, don't ever be afraid to plot the points on graph paper.

$$d = \sqrt{(x_2 - x_1)^2 + (y_2 - y_1)^2}$$

Distance of $\overline{MA}$: $MA = \sqrt{(4-1)^2 + (6-1)^2}$
$$= \sqrt{(3)^2 + (5)^2}$$
$$= \sqrt{34}$$

Distance of $\overline{AT}$: $AT = \sqrt{(7-4)^2 + (11-6)^2}$
$$= \sqrt{(3)^2 + (5)^2}$$
$$= \sqrt{34}$$

▶ Therefore, $\triangle MAT$ is an isosceles triangle because two sides of the triangle are congruent.

$$MA = AT = \sqrt{34}$$

In our next example, we will apply the distance formula to the properties of a rhombus.

▶ Parallelogram *GEOM* has vertices *G*(1,1), *E*(5,3), *O*(7,7), and *M*(3,5). Show that *GEOM* is a rhombus.

▶ A rhombus is a parallelogram in which all sides are congruent. We can find the distance between the endpoints of each side of the rhombus and show that they are equal in length.

$$d = \sqrt{(x_2 - x_1)^2 + (y_2 - y_1)^2}$$

Distance of $\overline{GE}$:  $GE = \sqrt{(5-1)^2 + (3-1)^2}$

$$= \sqrt{(4)^2 + (2)^2}$$
$$= \sqrt{20}$$

Distance of $\overline{EO}$:  $EO = \sqrt{(7-5)^2 + (7-3)^2}$

$$= \sqrt{(2)^2 + (4)^2}$$
$$= \sqrt{20}$$

Distance of $\overline{OM}$:  $OM = \sqrt{(3-7)^2 + (5-7)^2}$

$$= \sqrt{(-4)^2 + (-2)^2}$$
$$= \sqrt{20}$$

Distance of $\overline{MG}$:  $MG = \sqrt{(3-1)^2 + (5-1)^2}$

$$= \sqrt{(2)^2 + (4)^2}$$
$$= \sqrt{20}$$

▶ We can conclude that parallelogram *GEOM* is a rhombus because all four sides are equal.

## Midpoint Formula

A midpoint is a point on a line segment that divides the line segment into two congruent parts. It is **collinear** and **equidistant** from the endpoints of the line segment. The midpoint formula is useful when you need to find the center of a circle, the equation of a segment bisector, or the intersection point of the diagonals of a parallelogram.

**BTW**

*Think of the midpoint as the average of the two endpoints. An average of two numbers is found by adding the two numbers and then dividing it by 2.*

The following formula is used to find the midpoint. Take the average of the $x$ values and then take the average of the $y$ values. Remember to include the parentheses because the final result represents a point.

$$\text{Midpoint} = \left( \frac{x_2 + x_1}{2}, \frac{y_2 + y_1}{2} \right)$$

Let's try the following example to be sure we understand how to use the midpoint formula.

**EXAMPLE**

▶ Find the midpoint of a line segment with endpoints (3,7) and (5,–1).

▶ Use the midpoint formula to take the average of the $x$ values and the average of the $y$ values. Again, include the parentheses in the final answer to indicate that we are representing a point.

$$\begin{array}{cc} (x_1, y_1) & (x_2, y_2) \\ (3,7) & \text{and} \quad (5,-1) \end{array}$$

$$\begin{aligned} \text{Midpoint} &= \left( \frac{x_2 + x_1}{2}, \frac{y_2 + y_1}{2} \right) \\ &= \left( \frac{5+3}{2}, \frac{-1+7}{2} \right) \\ &= \left( \frac{8}{2}, \frac{6}{2} \right) \end{aligned}$$

▶ This means that the midpoint of this line segment is (4,3).

Our next example applies the midpoint formula to finding the center of a circle. Let's give it a try!

EXAMPLE

▶ What are the coordinates of the center of a circle whose diameter has endpoints at (4,10) and (8,6)?

▶ The center of a circle is the midpoint of the diameter.

Let $(x_1, y_1) = (4,10)$

Let $(x_2, y_2) = (8,6)$

$$\text{Midpoint} = \left( \frac{x_2 + x_1}{2}, \frac{y_2 + y_1}{2} \right)$$

$$= \left( \frac{4+8}{2}, \frac{10+6}{2} \right)$$

$$= \left( \frac{12}{2}, \frac{16}{2} \right) = (6,8)$$

▶ The center of the circle is (6,8).

Now let's look at an example where we are already given the midpoint of a line segment.

EXAMPLE

▶ $M(2,-4)$ is the midpoint of $\overline{ST}$. If $S$ has coordinates (5,1), what are the coordinates of point $T$?

▶ In this example, we are given the midpoint and one endpoint. We need to find the other endpoint. Because we know the $x$ value of the midpoint is 2 and the $y$ value of the midpoint is $-4$, we can set up the following:

$$\frac{x_2 + x_1}{2} = 2 \qquad \frac{y_2 + y_1}{2} = -4$$

▶ We can now substitute in the given $x$ and $y$ coordinates for point $S$:

$$\frac{5+x}{2} = 2 \qquad \frac{1+y}{2} = -4$$

▶ Placing a 1 in the denominator of the right-hand side of both equations will help you realize that you can cross multiply to solve for the variables.

$$\begin{array}{rr} \cancel{5}+x=4 & \cancel{1}+y=-8 \\ \underline{-\cancel{5} \quad -5} & \underline{-\cancel{1} \quad -1} \\ x=-1 & y=-9 \end{array}$$

▶ The other endpoint is $T(-1,-9)$.

Lastly, let's apply the midpoint formula toward the properties of a parallelogram.

▶ The vertices of parallelogram $MATH$ are represented by $M(2,3)$, $A(11,6)$, $T(10,9)$, and $H(1,6)$. At what point do the diagonals of the parallelogram intersect?

▶ The diagonals of a parallelogram bisect each other, which means they intersect at the same midpoint. Feel free to graph the vertices of the parallelogram and connect the diagonals to visually understand that property.

▶ To find the intersection point of the diagonals of a parallelogram, we need to find the midpoint of either diagonal.

$$\text{Midpoint} = \left( \frac{x_2+x_1}{2}, \frac{y_2+y_1}{2} \right)$$

$$\text{Midpoint of } \overline{MT} = \left( \frac{2+10}{2}, \frac{3+9}{2} \right) = (6,6)$$

$$\text{Midpoint of } \overline{AH} = \left( \frac{11+1}{2}, \frac{6+6}{2} \right) = (6,6)$$

▶ The diagonals intersect at $(6,6)$.

## Slope Formula

As you probably remember from algebra, the slope of a line, *m*, is a numerical value representing the direction and steepness of a line. In geometry, the slope formula is useful in showing properties of geometric figures containing sides that are parallel or perpendicular.

 **IRL**    Slope is considered in everyday situations when building roads, wheelchair ramps, and rooftops. Athletes such as skiers and golfers need to consider the slope of mountains and greens in their respective sports.

The slope formula represents the change in *y* values of two points divided by the change in *x* values of these two points (rise over run).

**BTW**

*Avoid the common mistake of putting the x's in the numerator of the formula and the y's in the denominator of the formula. Also, be sure to keep the order of the variables consistent when using the formula; otherwise, your slope will end up being the wrong sign.*

Slope formula:    $m = \dfrac{y_2 - y_1}{x_2 - x_1}$

Before we begin using the slope formula to prove different geometric figures, let's practice finding the slope between two points.

**EXAMPLE**

▶ What is the slope of a line segment whose endpoints are $H(-1,4)$ and $I(2,6)$?

$$m\overline{HI} = \frac{y_2 - y_1}{x_2 - x_1} = \frac{6 - 4}{2 - (-1)}$$

$$= \frac{2}{3}$$

It is important to realize that it does not matter which point you designate as $(x_1,y_1)$ or $(x_2,y_2)$. If you chose $H(-1,-4)$ as your $(x_2,y_2)$ instead of as your $(x_1,y_1)$, the numerator would have been negative and the denominator would have been negative. Because the quotient of two negative numbers is positive, the answer will remain the same.

$$m\overline{HI} = \frac{y_2-y_1}{x_2-x_1} = \frac{4-6}{(-1)-2}$$
$$= \frac{-2}{-3} = \frac{2}{3}$$

**EASY MISTAKE**

This would be incorrect:

$$m = \frac{4-6}{2-(-1)}$$

As you can see, the answer would come out as $-\frac{2}{3}$, which is incorrect.

We discussed earlier that slope is useful in determining whether line segments are parallel or perpendicular. Lines that are parallel either both rise or both fall at the same rate, which is why they never intersect. This means that **parallel lines have slopes that are equal**. Lines that are perpendicular move in different directions with the horizontal and vertical distance between two points being reversed. This means that **perpendicular lines have slopes that are negative reciprocals** of one another. If one line has a slope of $\frac{3}{4}$, it will be perpendicular to a line whose slope is $-\frac{4}{3}$.

Let's use the slope formula to show that the lines in the following example are parallel.

**EXAMPLE**

$\overline{AB}$ has endpoints $A(-2,3)$, and $B(0,7)$. $\overline{CD}$ has endpoints $C(5,-4)$ and $D(8,2)$. Show that $\overline{AB}$ is parallel to $\overline{CD}$.

Parallel lines have equal slopes. We can prove the line segments are parallel by using the slope formula to show that the slopes of the lines are equal.

$$m\overline{AB} = \frac{y_2 - y_1}{x_2 - x_1} = \frac{7 - 3}{0 - (-2)} = \frac{4}{2} \qquad m\overline{CD} = \frac{y_2 - y_1}{x_2 - x_1} = \frac{2 - (-4)}{8 - 5} = \frac{6}{3}$$

$$m\overline{AB} = \frac{2}{1} \qquad\qquad\qquad m\overline{CD} = \frac{2}{1}$$

▶ The slopes of $\overline{AB}$ and $\overline{CD}$ are equal, which means that $\overline{AB}$ is parallel to $\overline{CD}$.

The next example uses the slope formula to prove that two line segments are perpendicular.

**EXAMPLE**

▶ $\overline{ME}$ has endpoints $M(4,1)$ and $E(8,7)$. $\overline{DO}$ has endpoints $D(-4,2)$ and $O(-10,6)$. Show that $\overline{ME}$ and $\overline{DO}$ are perpendicular.

▶ Perpendicular lines have negative reciprocal slopes. We can prove the line segments are perpendicular by using the slope formula to show that the slopes are negative reciprocals.

$$m\overline{ME} = \frac{y_2 - y_1}{x_2 - x_1} = \frac{7 - 1}{8 - 4} = \frac{6}{4} \qquad m\overline{DO} = \frac{y_2 - y_1}{x_2 - x_1} = \frac{6 - 2}{-10 - (-4)} = \frac{4}{-6}$$

$$m\overline{ME} = \frac{3}{2} \qquad\qquad\qquad m\overline{DO} = -\frac{2}{3}$$

▶ Notice that the slopes of $\overline{ME}$ and $\overline{DO}$ are opposite signs and that the fractions are reciprocals of one another. $\overline{ME}$ and $\overline{DO}$ are perpendicular to one another because their slopes are negative reciprocals.

Now that we understand how to use the slope formula to prove lines parallel or perpendicular, let's apply the concept of slope to the properties of a rectangle in the following example.

EXAMPLE

▶ The vertices of parallelogram *MATH* are represented by *M*(2,3), *A*(11,6), *T*(10,9), and *H*(1,6). Prove that *MATH* is a rectangle.

▶ You might find it helpful to graph *MATH*.

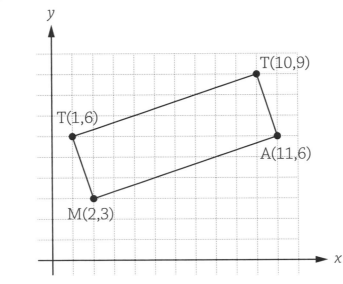

▶ Normally, to prove a shape is a rectangle we first need to prove it is a parallelogram and then show that it has a right angle. However, this example already stated that *MATH* is a parallelogram, which means all we need to do is show that this is a rectangle by showing that it contains one right angle.

▶ What forms right angles? Perpendicular lines! This means to show that *MATH* has a right angle we need to pick any two adjacent sides of the parallelogram and show they are perpendicular to one another. We know that perpendicular lines have slopes that are negative reciprocals of one another. We can show the adjacent sides of the parallelogram are perpendicular by showing that their slopes are negative reciprocals.

$$m\overline{MA} = \frac{y_2 - y_1}{x_2 - x_1} = \frac{6-3}{11-2} = \frac{3}{9} = \frac{1}{3}$$

$$m\overline{AT} = \frac{y_2 - y_1}{x_2 - x_1} = \frac{6-9}{11-10} = -\frac{3}{1}$$

▶ $\overline{MA}$ is perpendicular to $\overline{AT}$ because their slopes are negative reciprocals. Because perpendicular lines form right angles, angle $A$ must be a right angle. We can now conclude that $MATH$ is a rectangle because it is a parallelogram that contains a right angle.

If you truly understand the properties for the different parallelograms we learned in the last chapter, you might have already made the connection that we can prove the different properties using the distance, midpoint, and slope formulas. The distance formula proves that segments are congruent, the midpoint formula proves that segments bisect each other, and the slope formula proves that segments are parallel or perpendicular.

Let's use coordinate geometry formulas to show some of the properties of a square. As a reminder, you can see the list of the special properties of a square below.

Property 1    **A square is a parallelogram.**

Property 2    **All four sides of a square are congruent.**

Property 3    **Diagonals of a square are perpendicular.**

Property 4    **Diagonals of a square bisect each other.**

Property 5    **Adjacent sides of a square form a right angle.**

Property 6    **Diagonals of a square are congruent.**

We have already practiced showing quadrilaterals are parallelograms, which means we will begin by showing the second property using coordinate geometry.

▶ Property 2: All four sides of a square are congruent.

▶ Square $WXYZ$ has the vertices $W(5,1)$, $X(7,2)$, $Y(6,4)$, and $Z(4,3)$. Show that all four sides of the square are congruent.

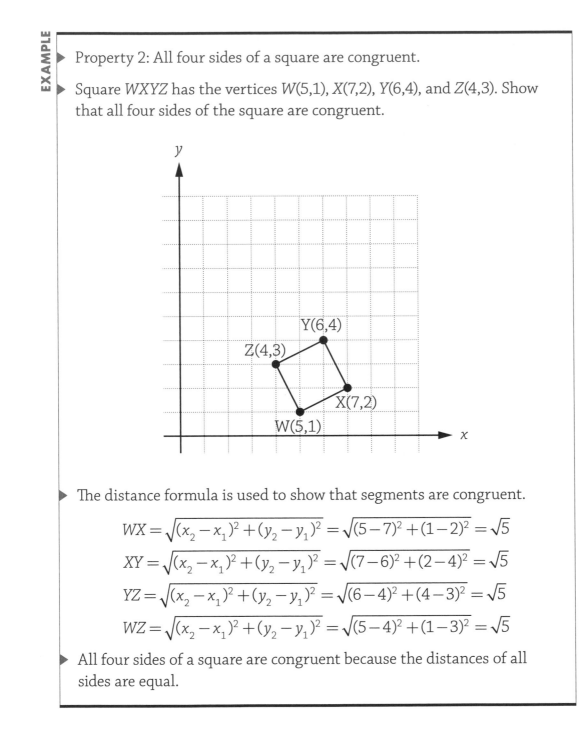

▶ The distance formula is used to show that segments are congruent.

$$WX = \sqrt{(x_2 - x_1)^2 + (y_2 - y_1)^2} = \sqrt{(5-7)^2 + (1-2)^2} = \sqrt{5}$$

$$XY = \sqrt{(x_2 - x_1)^2 + (y_2 - y_1)^2} = \sqrt{(7-6)^2 + (2-4)^2} = \sqrt{5}$$

$$YZ = \sqrt{(x_2 - x_1)^2 + (y_2 - y_1)^2} = \sqrt{(6-4)^2 + (4-3)^2} = \sqrt{5}$$

$$WZ = \sqrt{(x_2 - x_1)^2 + (y_2 - y_1)^2} = \sqrt{(5-4)^2 + (1-3)^2} = \sqrt{5}$$

▶ All four sides of a square are congruent because the distances of all sides are equal.

EXAMPLE

▶ Property 3: The diagonals of a square are perpendicular.

▶ Square *WXYZ* has the vertices *W*(5,1), *X*(7,2), *Y*(6,4), and *Z*(4,3). Show that the diagonals of the square are perpendicular.

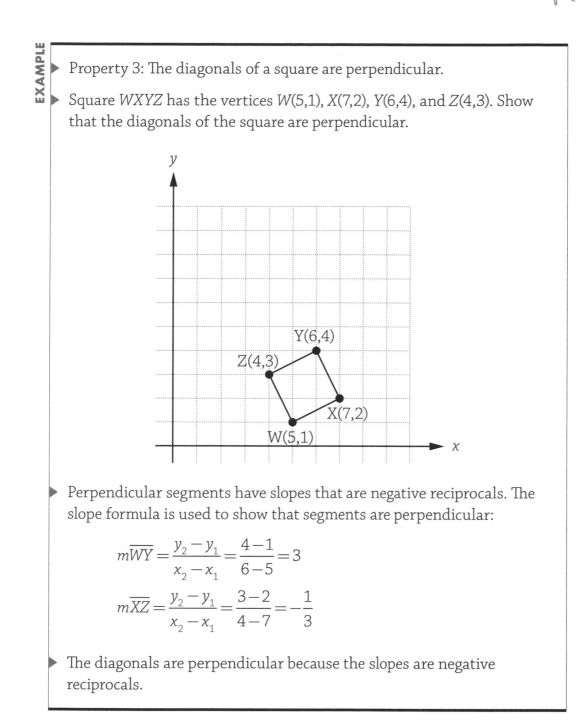

▶ Perpendicular segments have slopes that are negative reciprocals. The slope formula is used to show that segments are perpendicular:

$$m\overline{WY} = \frac{y_2 - y_1}{x_2 - x_1} = \frac{4-1}{6-5} = 3$$

$$m\overline{XZ} = \frac{y_2 - y_1}{x_2 - x_1} = \frac{3-2}{4-7} = -\frac{1}{3}$$

▶ The diagonals are perpendicular because the slopes are negative reciprocals.

▶ Property 4: The diagonals of a square bisect each other.

▶ Square $WXYZ$ has the vertices $W(5,1)$, $X(7,2)$, $Y(6,4)$, and $Z(4,3)$. Show that the diagonals of the square bisect each other.

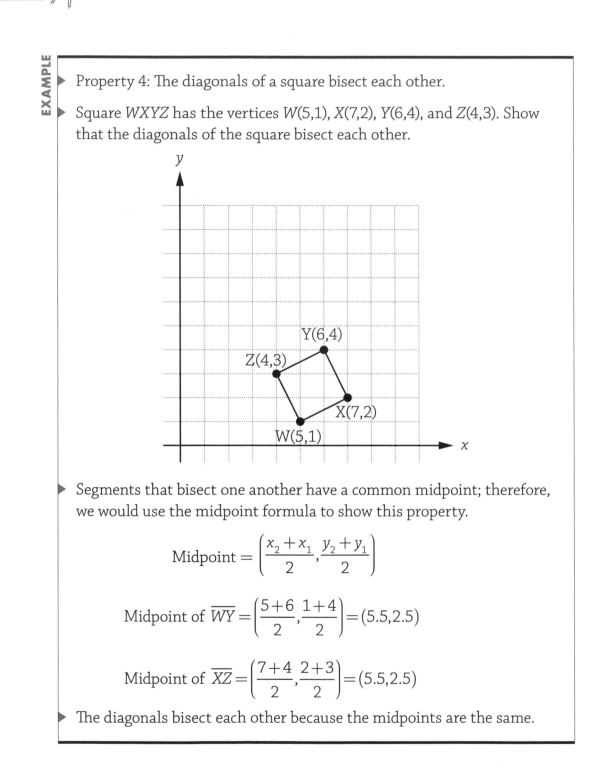

▶ Segments that bisect one another have a common midpoint; therefore, we would use the midpoint formula to show this property.

$$\text{Midpoint} = \left(\frac{x_2 + x_1}{2}, \frac{y_2 + y_1}{2}\right)$$

$$\text{Midpoint of } \overline{WY} = \left(\frac{5+6}{2}, \frac{1+4}{2}\right) = (5.5, 2.5)$$

$$\text{Midpoint of } \overline{XZ} = \left(\frac{7+4}{2}, \frac{2+3}{2}\right) = (5.5, 2.5)$$

▶ The diagonals bisect each other because the midpoints are the same.

**EXAMPLE**

Property 5: The adjacent sides of a square form a right angle.

Square *WXYZ* has the vertices *W*(5,1), *X*(7,2), *Y*(6,4), and *Z*(4,3). Show that the adjacent sides of the square form a right angle. You only have to do this with one pair of adjacent sides.

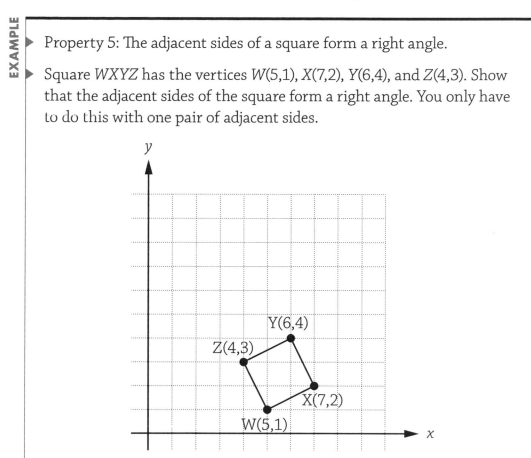

Perpendicular lines form right angles. We will use the slope formula to show that the adjacent sides are perpendicular and, therefore, form right angles.

$$m\overline{WX} = \frac{y_2 - y_1}{x_2 - x_1} = \frac{2-1}{7-5} = \frac{1}{2}$$

$$m\overline{XY} = \frac{y_2 - y_1}{x_2 - x_1} = \frac{4-2}{6-7} = -\frac{2}{1}$$

The slopes of the adjacent sides are negative reciprocals, which means the adjacent sides are perpendicular. Because perpendicular lines form right angles, $\angle WXY$ is a right angle.

▶ Property 6: The diagonals of a square are congruent.

▶ Square $WXYZ$ has the vertices $W(5,1)$, $X(7,2)$, $Y(6,4)$, and $Z(4,3)$. Show that the diagonals of the square are congruent.

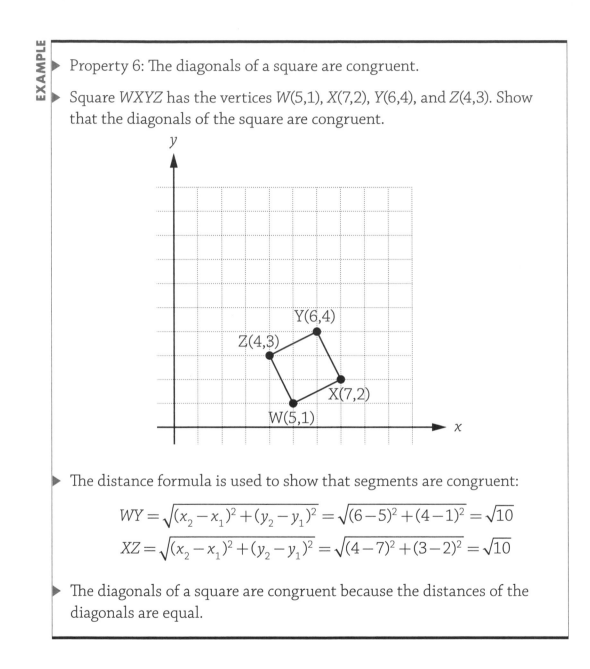

▶ The distance formula is used to show that segments are congruent:

$$WY = \sqrt{(x_2 - x_1)^2 + (y_2 - y_1)^2} = \sqrt{(6-5)^2 + (4-1)^2} = \sqrt{10}$$
$$XZ = \sqrt{(x_2 - x_1)^2 + (y_2 - y_1)^2} = \sqrt{(4-7)^2 + (3-2)^2} = \sqrt{10}$$

▶ The diagonals of a square are congruent because the distances of the diagonals are equal.

# Writing the Equations of Parallel and Perpendicular Lines

We have already learned that parallel lines have equal slopes and perpendicular lines have slopes that are negative reciprocals; therefore, we should be able to look at the equations of lines and determine if they are parallel or perpendicular. To do this, we must have the equations of the line written in **slope-intercept form**, $y = mx + b$, with $m$ (the coefficient of the $x$) representing the slope of the line.

Consider the lines $y = 3x + 5$ and $y = 3x - 7$. These lines have the same slope, which means the lines are parallel.

Now consider the lines $y = 3x - 1$ and $y = -\dfrac{1}{3}x + 4$. These lines have slopes that are negative reciprocals, which means the lines are perpendicular.

To write an equation of a line that is parallel or perpendicular to a given line, we use the **point-slope formula** of a line: $y - y_1 = m(x - x_1)$. To use this formula, we need the slope of the line, $m$, and a point on the line $(x_1, y_1)$.

The following example shows how to write the equation of a line parallel to a given line and passing through a given point.

**EXAMPLE**

▶ Write the equation of a line that is parallel to $4x - 6 = 2y$ and passes through the point (8,10).

▶ We first need to find the slope of the given line. To do this, we must convert the line into slope-intercept form by solving for $y$:

$$2y = 4x - 6$$
$$\frac{2y}{2} = \frac{4x - 6}{2}$$
$$y = 2x - 3$$

▶ The slope of the line is 2. Therefore, the slope of any line parallel to this line must also be 2 because parallel lines have equal slopes. We can now substitute the slope and the point (8,10) into the point-slope form of a line to get the equation of the new line.

Let $(x_1, y_1) = (8,10)$

Let $m = 2$

$$y - y_1 = m(x - x_1)$$
$$y - 10 = 2(x - 8)$$
$$y - \cancel{10} = 2x - 16$$
$$\underline{+\cancel{10} \qquad +10}$$
$$y = 2x - 6$$

▶ Therefore the equation of the line parallel to $4x - 6 = 2y$ and passing through the point (8,10) is $y = 2x - 6$.

Next, we will look at a similar problem but examine what happens when you want a line that is perpendicular to the given line.

EXAMPLE

▶ What is the equation of a line that is perpendicular to $3x + 4y = 2$ and passes through the point (12,1)?

▶ Let's begin by converting the equation into slope-intercept form by solving for $y$:

$$3x + 4y = 2$$
$$4y = -3x + 2$$
$$\frac{4y}{4} = \frac{-3x + 2}{4}$$
$$y = -\frac{3}{4}x + \frac{1}{2}$$

▶ The slope of the given line is $-\frac{3}{4}$. Therefore, the slope of any line perpendicular to this line has to be the negative reciprocal of $-\frac{3}{4}$, which is $\frac{4}{3}$. We can substitute the slope of $\frac{4}{3}$ and the point (12,1) into the point-slope form of a line to get the equation of the new line.

$$\text{Let } (x_1, y_1) = (12,1)$$

$$\text{Let } m = \frac{4}{3}.$$

$$y - y_1 = m(x - x_1)$$

$$y - 1 = \frac{4}{3}(x - 12)$$

$$y - \cancel{1} = \frac{4}{3}x - 16$$

$$\underline{+\cancel{1} \qquad\quad +1}$$

$$y = \frac{4}{3}x - 15$$

▶ Therefore, the equation of the line that is perpendicular to $3x + 4y = 2$ and passes through the point (12,1) is $y = \frac{4}{3}x - 15$.

## Partitioning a Line Segment

The midpoint formula locates the point that divides a line segment into two congruent parts, giving the two smaller segments a 1:1 ratio. Not everything has to be divided equally.

Take a pretzel stick, for example. You might want to give your sibling the smaller piece of the pretzel stick and keep a bigger piece for yourself. Don't break it at the midpoint! Pick a location that will give you a piece two times as big. To do this you would divide the pretzel into a 2:1 ratio, as shown in the following diagram.

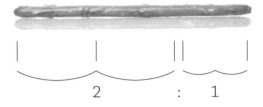

2 : 1

Dividing the pretzel in a 2:1 ratio means that you will be sectioning it into three congruent pieces and taking $\frac{2}{3}$ of it. In geometry, we call this *partitioning a line segment*.

The example below shows how to locate the point that divides a line segment into a given ratio.

**EXAMPLE**

▶ A(–5,2) and B(7,10) are the endpoints of $\overline{AB}$. Find the point P on $\overline{AB}$ such that the ratio of $\overline{AP}$ to $\overline{PB}$ is 3:1.

▶ Graphing $\overline{AB}$ may make it easier for you to visualize the steps for dividing the segment into the given ratio.

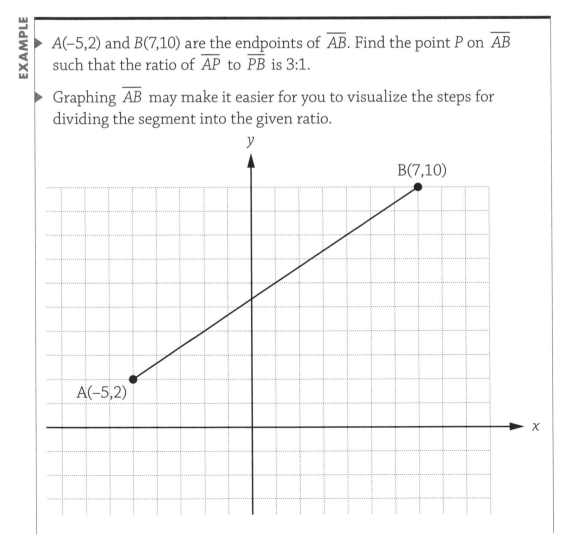

▶ The ratio 3:1 means that the line segment is being divided in such a way that $\overline{AP}$ contains three congruent parts and $\overline{PB}$ contains one congruent part for a total of four congruent parts. We are looking to find the location of point $P$ such that $\overline{AP}$ is $\frac{3}{4}$ of the total distance of $\overline{AB}$.

$P$ is a point, which means it consists of an $x$ and $y$ value. Let's locate the $x$ value first.

▶ **Finding the x value**

▶ We want the horizontal distance from point $A$ to point $P$ to be $\frac{3}{4}$ of the horizontal distance from $A$ to $B$. The horizontal distance is the difference in the $x$ values of the endpoints: $x_2 - x_1$.

$$7 - (-5) = 12$$

▶ The horizontal distance from point $A$ to point $P$ must be $\frac{3}{4}$ of 12: $\frac{3}{4}(12) = 9$. We need the $x$ value for point $P$ to be nine units to the right of the $x$ value for point $A$:

$$-5 + 9 = 4$$

▶ The $x$ value for point $P$ is 4. Now it is time to find the $y$ value of point $P$.

▶ **Finding the y value**

▶ We want the vertical distance from point $A$ to point $P$ to be $\frac{3}{4}$ of the vertical distance from point $A$ to point $B$. The vertical distance is the difference in the $y$ values of the two endpoints: $y_2 - y_1$.

$$10 - 2 = 8$$

▶ The vertical distance from point $A$ to point $P$ must be $\frac{3}{4}$ of 8: $\frac{3}{4}(8) = 6$. We need the $y$ value for point $P$ to be six units above the $y$ value for point $A$:

$$2 + 6 = 8$$

▶ The $y$ value for point $P$ is 8. Therefore $P(4,8)$ is the location that will divide $\overline{AB}$ into a 3:1 ratio.

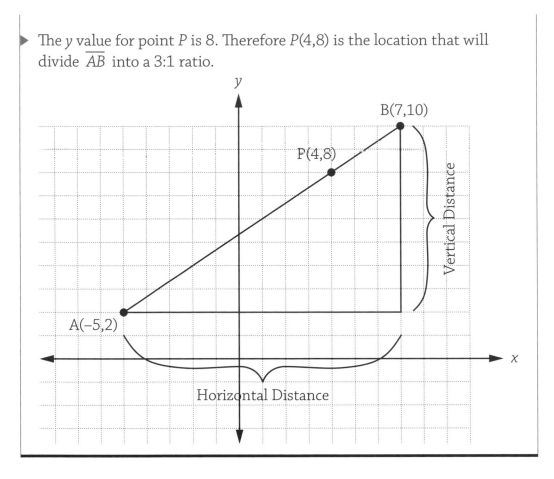

The next example is similar but requires you to partition the line segment in a different ratio algebraically.

**BTW**

*There is a difference between 3:1 and 1:3. A 1:3 ratio is also divided into four congruent sections, but the first segment is ¼ of the total distance.*

**EXAMPLE**

▶ Given: $J(3,-1)$ and $N(17,20)$. Algebraically, find the point $O$ on $\overline{JN}$ such that the ratio of $\overline{JO}$ to $\overline{ON}$ is 2:5.

▶ The ratio 2:5 means that the line segment is being divided in such a way that $\overline{JO}$ contains two congruent parts and $\overline{ON}$ contains five congruent parts for a total of seven congruent parts. We are looking to find the location of point $O$ such that $\overline{JO}$ is $\dfrac{2}{7}$ of the total distance of $\overline{JN}$. $O$ is a point, which means it consists of an $x$ and a $y$ value. Let's locate the $x$ value first.

► **Finding the $x$ value**

► We want the horizontal distance from point $J$ to point $O$ to be $\dfrac{2}{7}$ of the horizontal distance from point $J$ to point $N$. The horizontal distance is the difference in the $x$ values of the endpoints: $x_2 - x_1$.

$$17 - 3 = 14$$

► The horizontal distance from point $J$ to point $O$ must be $\dfrac{2}{7}$ of 14: $\dfrac{2}{7}(14) = 4$. We need the $x$ value for point $O$ to be four units to the right of the $x$ value for point $J$:

$$3 + 4 = 7$$

► The $x$ value for point $O$ is 7. Now it is time to find the $y$ value of point $O$.

► **Finding the $y$ value**

► We want the vertical distance from point $J$ to point $O$ to be $\dfrac{2}{7}$ of the vertical distance from point $J$ to point $N$. The vertical distance is the difference in the $y$ values of the two endpoints: $y_2 - y_1$.

$$20 - (-1) = 21$$

► The vertical distance from point $J$ to point $O$ must be $\dfrac{2}{7}$ of 21: $\dfrac{2}{7}(21) = 6$. We need the $y$ value for point $O$ to be six units above the $y$ value for point $J$.

$$-1 + 6 = 5$$

► The $y$ value for point $O$ is 5. Therefore $O(7,5)$ is the location that will divide $\overline{JN}$ into a 2:5 ratio.

## EXERCISES

### EXERCISE 9-1

*For problems 1–4, find the length of a line segment with the given endpoints.*

1. $A(4,15)$ and $B(4,25)$

2. $N(-2,14)$ and $P(11,14)$

3. $S(3,12)$ and $T(6,8)$

4. $H(-5,-1)$ and $K(3,1)$

### EXERCISE 9-2

*Quadrilateral BASE has vertices B(4,2), A(6,3), S(5,5), and E(3,4). Use this information to answer the following questions.*

1. Is $\overline{BA}$ parallel to $\overline{SE}$?

2. Are the diagonals $\overline{SB}$ and $\overline{AE}$ perpendicular to each other?

3. Are the consecutive sides $\overline{AS}$ and $\overline{SE}$ congruent?

4. Do the diagonals $\overline{SB}$ and $\overline{AE}$ bisect each other?

### EXERCISE 9-3

*Solve the following problems.*

1. Use the distance formula to determine if parallelogram $RHOM$ with vertices $R(3,-1)$, $H(7,1)$, $O(9,5)$, and $M(5,3)$ is a rhombus.

2. The endpoints of the diameter of a circle are $(1,18)$ and $(3,12)$. Find the center of the circle.

3. Find the midpoint of a line segment with endpoints $(17,-2)$ and $(4,-8)$.

4. $M(-2,10)$ is the midpoint of $\overline{AT}$. $A$ has coordinates $(4,25)$. Find the coordinates of $T$.

5. Find the slope of a line that passes through points $W(3,15)$ and $Y(-3,18)$.

6. Find the slope of a line that passes through points $C(1,-6)$ and $D(7,12)$.

7. What is the slope of a line that is parallel to $y = -2x - 1$?

8. What is the slope of a line that is perpendicular to $2y = 3x + 5$?

9. What is the equation of a line that is perpendicular to $8x + 2y = 10$ and passes through the point $(16,20)$?

10. What is the equation of a line that is parallel to $6x + 3y = 21$ and passes through the point $(10,10)$.

11. $P(-2,5)$ and $S(6,29)$ are the endpoints of $\overline{PS}$. Find the coordinates for point $A$ on $\overline{PS}$ such that the ratio of $\overline{PA}$ to $\overline{AS}$ is 3:1.

12. $C(5,11)$ and $D(8,17)$ are the endpoints of $\overline{CD}$. Find the coordinates of point $T$ on $\overline{CD}$ such that the ratio of $\overline{CT}$ to $\overline{DT}$ is 2:1.

Flashcard App

# Transformations

## MUST ⚡ KNOW

⚡ A transformation involves an object changing its location on the coordinate plane.

⚡ The rules for reflections, rotations, translations, and dilations determine where the image of a point or geometric figure will lie on the coordinate plane.

⚡ Transformations that are rigid motions preserve distance and angle measure.

⚡ A reflection "flips" an object over a point or line.

⚡ A rotation "spins" an object about a fixed point.

⚡ In a translation, an object "slides" across the coordinate plane.

⚡ In a dilation, an object "shrinks" or "stretches." A dilation preserves angle measure, *not* distance.

transformation means a change is occurring. We've all heard of the movie *Transformers*. Robots transform themselves into cars, planes, weapons, buildings, and other amazing objects! In geometry, transformations take a set of points or shapes and change their location on the coordinate plane. The four transformations that will be discussed in this chapter are reflections, translations, rotations, and dilations.

A rigid motion is when an object's size and shape remain the same after a transformation takes place. For example, if you were to simply rotate your phone on the table, the position would most likely be different, but the size and shape of the phone would remain the same. This rotation would be an example of a rigid motion. A rigid motion preserves the distance and angle measure of the shape.

## Reflections

When you look in the mirror, what do you see? A reflection of yourself! If you hold up your right hand when looking in the mirror, the image you see is the flipped version of yourself. In other words, the image is of you holding up your left hand. This is why we think of reflections as a flipping of an object over a line or point. In this book, we will learn how to reflect over different lines and points.

**BTW**

*The original point (the point before the transformation) is known as the pre-image. The term used for the point after the transformation takes place is known as the image.*

### Reflection Over the Y-Axis

The following diagram shows $\triangle ABC$ being reflected over the $y$-axis and resulting in $\triangle A'B'C'$.

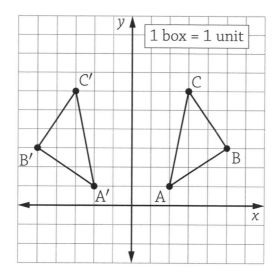

The pre-image and image points are as follows:

$$A(2,1) \rightarrow A'(-2,1)$$
$$B(5,3) \rightarrow B'(-5,3)$$
$$C(3,6) \rightarrow C'(-3,6)$$

Did you notice that the $x$ coordinates changed sign for the image points after a reflection over the $y$-axis? That's why the rule for reflecting over the $y$-axis is to negate the $x$ coordinate.

$$P(x,y) \xrightarrow[r_{y-axis}]{} P'(-x,y)$$

The image of the original triangle maintained its size and shape when it was reflected. Therefore, a reflection is a **rigid motion** because it preserves distance and angle measure. However, the order of the points was not maintained. When naming the original triangle, the points $A$, $B$, and $C$ are in counterclockwise order. The image $A'$, $B'$, $C'$ is in clockwise order. This means that reflections do not preserve **orientation**. Therefore, a line reflection is a rigid motion but considered an *opposite isometry* (an *isometry* is another term stating that distance is preserved).

## Reflection Over the X-Axis

The following diagram shows $\triangle ABC$ being reflected over the x-axis, resulting in $\triangle A'B'C'$.

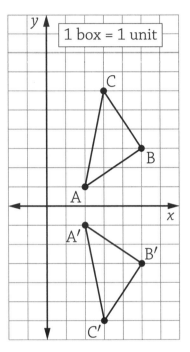

The pre-image and image points are as follows:

$$A(2,1) \rightarrow A'(2,-1)$$
$$B(5,3) \rightarrow B'(5,-3)$$
$$C(3,6) \rightarrow C'(3,-6)$$

Note that the y coordinates changed sign for the image points after a reflection over the x-axis. That's why the rule for reflecting over the x-axis is to negate the y coordinate.

$$P(x,y) \xrightarrow{\ r_{x-axis}\ } P'(x,-y)$$

## Reflection Over the Line y = x

The following diagram shows $\triangle ABC$ being reflected over the line $y = x$ and resulting in $\triangle A'B'C'$.

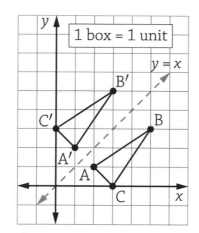

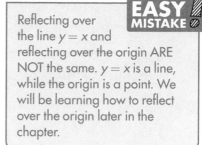

Reflecting over the line $y = x$ and reflecting over the origin ARE NOT the same. $y = x$ is a line, while the origin is a point. We will be learning how to reflect over the origin later in the chapter.

The pre-image and image points are as follows:

$$A(2,1) \rightarrow A'(1,2)$$
$$B(5,3) \rightarrow B'(3,5)$$
$$C(3,0) \rightarrow C'(0,3)$$

Did you notice that the $x$ and $y$ coordinates changed places after a reflection over the line $y = x$? That's why the rule for reflecting over the line $y = x$ is to switch the $x$ and $y$ coordinates.

$$P(x,y) \underset{r_{y=x}}{\longrightarrow} P'(y,x)$$

## Reflecting a Point Over Horizontal and Vertical Lines

The following diagram shows point $P(5,2)$ being reflected over the line $x = 2$, resulting in $P'(-1,2)$. Point $P$ is three units to the right of the line $x = 2$, which is why the image $P'$ needs to be three units to the left of the line $x = 2$.

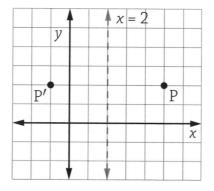

Let's apply that same concept toward reflecting a point over a horizontal line in the following example.

▶ What is the image of point $P(5,2)$ after a reflection over the line $y = 4$?

▶ As you can see in the following diagram, point $P$ is two units below the line $y = 4$. The image $P'$ must be two units above the line $y = 4$. Therefore, the image $P'$ would be $(5,6)$.

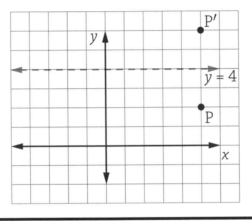

## EXTRA HELP

Let's apply all of the reflection rules we have reviewed so far towards the point $A(-2,5)$ WITHOUT graphing it.

1. Reflect point $A$ over the $x$-axis:
   When we reflect a point over the $x$-axis we negate the $y$ value; therefore, its image is $(-2,-5)$.

2. Reflect point $A$ over the $y$-axis:
   When we reflect a point over the $y$-axis we negate the $x$ value; therefore, its image is $(2,5)$.

3. Reflect point $A$ over the line $y = x$:
   When we reflect a point over the line $y = x$ we switch the $x$ and $y$ coordinates; therefore, its image is $(5,-2)$.

4. Reflect point $A$ over the line $x = 3$:

The $x$ value for point $A$ is five units lower than $x = 3$, so we will move five units higher to get the point (8,5).

5. Reflect point $A$ over the line $y = 4$:
   The $y$ value for point $A$ is 1 unit higher than $y = 4$ so we will move one unit lower to get the point (−2,3).

An **invariant point** can be described as a point that does not change its position when a transformation is applied. Essentially, the fixed point is mapped onto itself. The pre-image and image are in the exact same location. For example, when the point (7,1) is reflected over the line $x = 7$, the image remains at (7,1). This is because the point (7,1) lands on the line $x = 7$.

The following example requires you to determine the line of reflection that would map a geometric figure onto itself.

**EXAMPLE**

Square $ABCD$ is graphed in the following diagram.

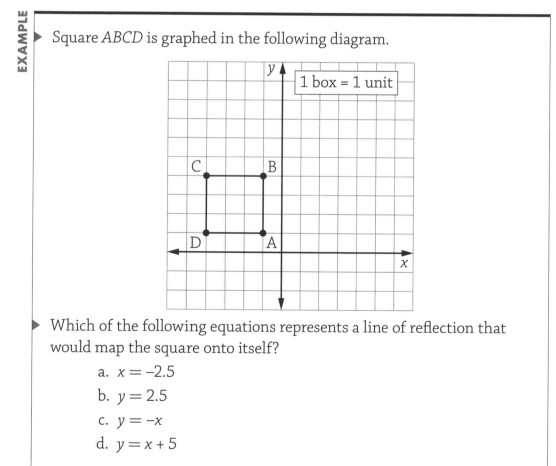

Which of the following equations represents a line of reflection that would map the square onto itself?

a. $x = -2.5$

b. $y = 2.5$

c. $y = -x$

d. $y = x + 5$

▶ All of the above! A common mistake is to think only $A$ and $B$. If you look at the graphs of all the lines, you can see that $C$ and $D$ are the diagonals of the square and therefore also would make the square map onto itself.

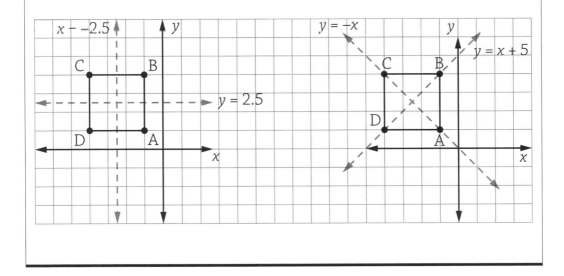

## Reflecting a Point Over an Oblique Line

Reflecting a point over an oblique (diagonal) line is more challenging because there are no rules to rely on as there were when we reflected over the $x$-axis, $y$-axis, and the line $y = x$. We cannot simply count boxes like we did when we reflected over a horizontal or vertical line. We must rely on our knowledge of slope, midpoint, and writing the equation of lines. The following example will give you a clear explanation of the steps needed to reflect a point over an oblique line.

EXAMPLE

▶ What is the image when point $P(0,7)$ is reflected over the line $y = 2x + 2$?

▶ Plot the point and graph the line of reflection.

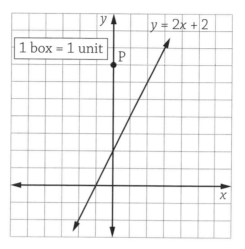

▶ We need to find the equation of the line perpendicular to $y = 2x + 2$ that passes through point $P$.

▶ The slope of the reflection line is 2. Because perpendicular lines have negative reciprocal slopes, the slope of the perpendicular line must be $-\frac{1}{2}$.

▶ The pre-image point is $(0,7)$. Using the point-slope form of a line, we can get the equation of the perpendicular line that passes through the pre-image point:

$$(x_1, y_1) = (0,7)$$

$$m = -\frac{1}{2}$$

$$y - y_1 = m(x - x_1)$$

$$y - 7 = -\frac{1}{2}(x - 0)$$

$$y - \cancel{7} = -\frac{1}{2}x$$

$$\underline{+\cancel{7} \quad +7}$$

$$y = -\frac{1}{2}x + 7$$

▶ This is represented as a dotted line in the following diagram.

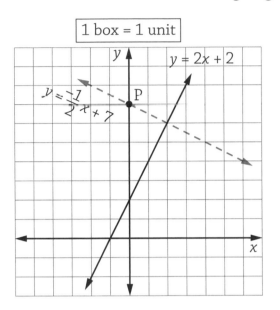

1 box = 1 unit

▶ Find the point where the two lines intersect by setting the equations of the lines equal, solving for $x$, and then substituting $x$ into the equation of the line to find $y$.

$$2x + \cancel{2} = -\cancel{\frac{1}{2}}x + 7$$

$$+\frac{1}{2}x - \cancel{2} \quad +\cancel{\frac{1}{2}}x - 2$$

$$\frac{\cancel{2.5}x}{\cancel{2.5}} = \frac{5}{2.5}$$

$$x = 2$$

$$y = 2x + 2$$

$$y = 2(2) + 2$$

$$y = 6$$

▶ The point of intersection is (2,6).

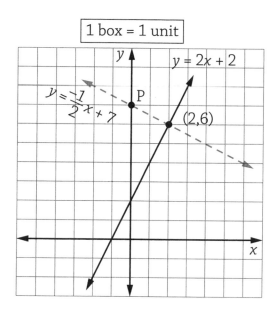

▶ (2,6) represents the midpoint of $P(0,7)$ and $P'$.

▶ To find $P'$: $P(0,7) \Rightarrow (2,6) \Rightarrow P'(x,y)$

▶ Finding the $x$ value: The $x$ value went from 0 to 2 to get to the midpoint. That means it went two units to the right. To get to the final location, go two more units to the right and arrive at $x = 4$.

$$P(0,\ )\underset{+2}{\rightarrow}(2,\ )\underset{+2}{\rightarrow}(4,\ )$$

▶ Finding the $y$ value: The $y$ value went from 7 to 6 to get to the midpoint. That means it went down one unit. To get to the final location, go down one more unit and arrive at $y = 5$.

$$P(\ ,7)\underset{-1}{\rightarrow}(\ ,6)\underset{-1}{\rightarrow}(\ ,5)$$

▶ Therefore, the image is $P'(4,5)$.

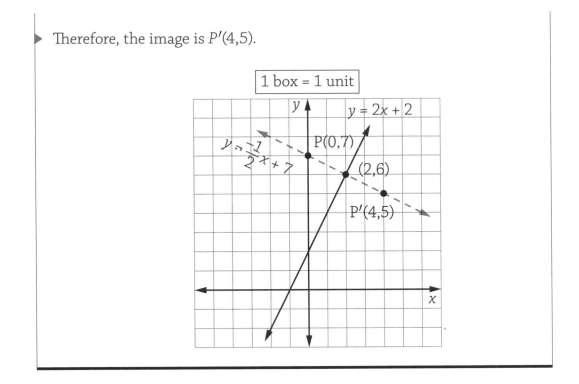

1 box = 1 unit

## Finding the Equation for the Line of Reflection

What if we are given the pre-image point and the image point but not the line of reflection? Would you be able to come up with the equation for the line of reflection? Give it a try with this next example.

**EXAMPLE**

▶ When point $A(-4,-2)$ is reflected over line $\ell$, the image is $A'(0,-6)$. Write the equation for line $\ell$.

▶ The reflection line is the perpendicular bisector of the line segment whose endpoints are the pre-image and image points.

▶ Find the slope of the line segment whose endpoints are $A$ and $A'$.

$$m = \frac{y_2 - y_1}{x_2 - x_1} = \frac{-6 - (-2)}{0 - (-4)} = \frac{-4}{4} = -1$$

▶ Find the midpoint of $\overline{AA'}$.

$$\left(\frac{x_2+x_1}{2}, \frac{y_2+y_1}{2}\right) = \left(\frac{-4+0}{2}, \frac{-2+(-6)}{2}\right) = (-2, -4)$$

▶ Write the equation for the line that passes through the midpoint of $\overline{AA'}$ and is also perpendicular to $\overline{AA'}$.

▶ Perpendicular lines have negative reciprocal slopes. Because the slope of $\overline{AA'}$ was –1, the slope of the reflection line must be 1. We can now use the point slope form of a line to get the equation for the reflection line.

$$\text{Let } (x_1, y_1) = (-2, -4)$$
$$m = 1$$
$$y - y_1 = m(x - x_1)$$
$$y - (-4) = 1(x - (-2))$$
$$y + 4 = 1(x + 2)$$
$$y + \cancel{4} = x + 2$$
$$\underline{-\cancel{4} \qquad -4}$$
$$y = x - 2$$

▶ The equation for the line of reflection is $y = x - 2$ and can be seen in the following diagram.

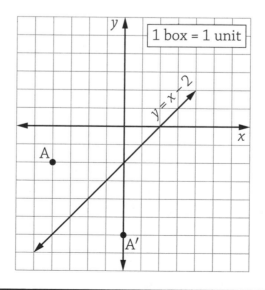

## Point Reflections

Reflecting a point over the origin is the most common type of point reflection. We use the symbol $R_0$ to indicate a point reflection over the origin. It is equivalent to rotating a shape $180°$ with the center at the origin. We will discuss rotations in the next part of the chapter. The following diagram shows $A'$, the image of point $A$, after a reflection over the origin.

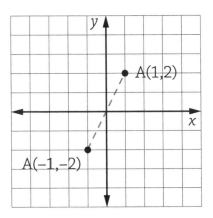

Rule: $A(x,y) \underset{R_o}{\longrightarrow} A'(-x,-y)$

# Rotations

A rotation is a transformation that spins a figure (clockwise or counterclockwise) about a fixed point called the **center of rotation**. Rotations preserve size and shape, which makes them a rigid motion transformation. We can also categorize a rotation as a *direct* isometry because it preserves distance and orientation. A counterclockwise rotation is represented by a positive degree of rotation, and a clockwise rotation is represented by a negative degree of rotation.

The following diagram shows $\triangle ABC$ being rotated $90°$ with the center of rotation at the origin.

The pre-image and image points are as follows:

$$A(2,1) \rightarrow A'(-1,2)$$

$$B(5,1) \rightarrow B'(-1,5)$$

$$C(5,3) \rightarrow C'(-3,5)$$

Keep in mind that the $y$ coordinate changed sign and then switched places with the $x$ coordinate.

Rule: $(x,y) \underset{R_{90°}}{\rightarrow} (-y,x)$

## Summary of Rules for Rotations with the Center of Rotation at the Origin

$$(x,y) \underset{R_{90°}}{\rightarrow} (-y,x)$$

$$(x,y) \underset{R_{180°}}{\rightarrow} (-x,-y)$$

$$(x,y) \underset{R_{270°}}{\rightarrow} (y,-x)$$

$$(x,y) \underset{R_{360°}}{\rightarrow} (x,y)$$

**EASY MISTAKE**

If the degree of rotation is negative it is not telling you to negate the coordinates. It is telling you to rotate in a clockwise direction. Since our rotation rules are for rotating in a counterclockwise direction (positive degree), add 360 degrees to the negative degree measure to turn it into a positive degree measure. For example rotating $-90$ degrees is equivalent to rotating $-90 + 360 = 270$ degrees.

**BTW**

When we rotate a point any multiple of $90°$ around the origin, we just use the rule $(x, y) \underset{R_{90°}}{\rightarrow} (-y, x)$ multiple times. For example, to rotate a point $270°$ around the origin, we complete the rule $(x, y) \underset{R_{90°}}{\rightarrow} (-y, x)$ three times.

### Rotation with Center of Rotation *Not* at the Origin

What if the center of rotation is not at the origin? If so, we cannot rely on the four rules just noted. The following example practices this skill.

**EXAMPLE**

▶ Rotate $\overline{AB}$ 270°, where $A(-5,3)$ and $B(-3,2)$ with the center of rotation at $C(1,-2)$.

▶ Graph $\overline{AB}$ and the center of rotation $C(4,-2)$.

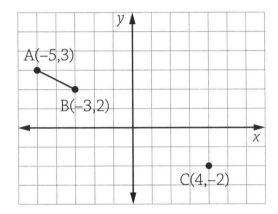

▶ Make $C(4,-2)$ the position of the "new origin." In other words, pretend that $(4,-2)$ is the origin.

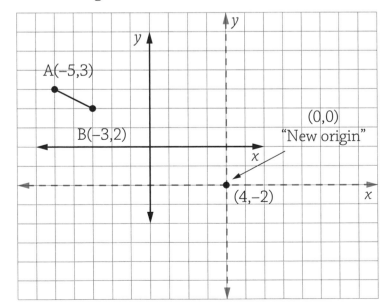

▶ If $C(4,-2)$ represents the "new origin," then $A(-5,3)$ is now at $(-9,5)$. If $C(4,-2)$ represents the "new origin," then $B(-3,2)$ is now at $(-7,4)$.

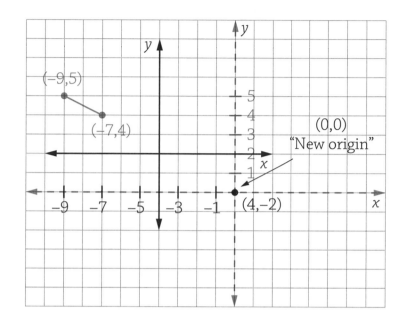

▶ Use the rule written for $R_{270°}$ with center at the origin.

$$(x,y) \underset{R_{270°}}{\longrightarrow} (y,-x)$$

$$(-9,5) \underset{R_{270°}}{\longrightarrow} (5,9)$$

$$(-7,4) \underset{R_{270°}}{\longrightarrow} (4,7)$$

▶ The point $(5,9)$ is relative to the "new origin." Now locate what $(5,9)$ really is with the origin in its normal location. It really is $A'(9,7)$.

▶ Do the same with (4,7). Figure out what that point really is with the origin in its normal location. It really is $B'$ (8,5).

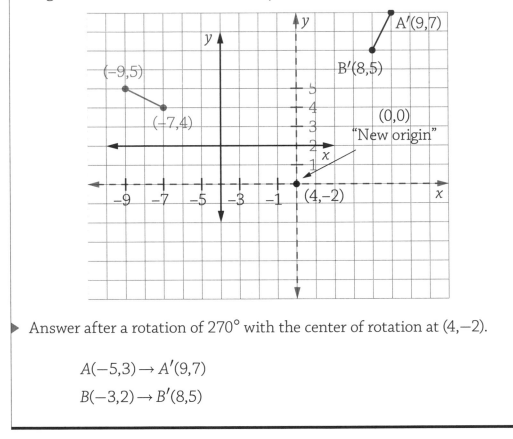

▶ Answer after a rotation of 270° with the center of rotation at (4,−2).

$$A(-5,3) \rightarrow A'(9,7)$$
$$B(-3,2) \rightarrow B'(8,5)$$

# Translations

A **translation** is defined as a shift or slide. It can be represented with a vector that shows the amount the pre-image needs to move and in what direction. Translations are considered a rigid motion because they preserve distance and angle measure. They are also direct isometries because they preserve orientation. You probably didn't realize it, but you already performed a translation in this chapter. Remember in the previous section when we were rotating a point around a center that wasn't the origin? We made (4,−2) the "new origin." We translated the origin four units to the right and two units down. Written mathematically, this would look like: $(0,0) \underset{T_{4,-2}}{\rightarrow} (4,-2)$.

Rule: $(x, y) \underset{T_{a,b}}{\rightarrow} (x+a, y+b)$

Let's practice this translation rule in the next example.

**EXAMPLE**

▶ The coordinates of $\overline{AM}$ are $A(-1,8)$ and $M(2,5)$. Find the coordinates of $\overline{A'M'}$ under the translation $T_{3,-5}$.

$$T_{3,-5} \Rightarrow (x+3, y-5)$$
$$A(-1,8) \underset{T_{3,-5}}{\rightarrow} A'(2,3)$$
$$M(2,5) \underset{T_{3,-5}}{\rightarrow} M'(5,0)$$

Now let's look at a problem that gives both the pre-image and image. Your job is to figure out the translation that took place.

**EXAMPLE**

▶ Find the rule for the translation that maps $A(7,-6)$ to $A'(12,-13)$.

▶ Compare the $x$ values from the pre-image to the image and then compare the $y$ values from the pre-image to the image to find the rule for the translation.

$$7 \overset{+5}{\rightarrow} 12 \text{ and } -6 \overset{-7}{\rightarrow} -13.$$

▶ Therefore, the translation is $T_{5,-7}$.

The next example requires you to find the translation rule from a graph.

▶ Using the following diagram, find the translation that maps
$\triangle TRA$ to $\triangle T'R'A'$.

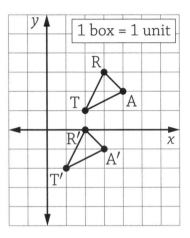

▶ You can do this by comparing the coordinates as we did in the previous
example, or you can actually count the boxes on the graph. Put your
pencil on point $T$. Now count one unit to the left and three units down.
Your pencil will now land on $T'$. Do the same with $R$ to $R'$ and $A$ to $A'$.
Each pre-image went one left and three down to arrive at its image.
Therefore, the translation is $T_{-1,-3}$.

## Dilation

A dilation is *not* a rigid motion. It is a transformation that does *not* preserve
distance. Its image is *not* congruent to the pre-image. Dilations do, however,
preserve angle measure. This means that even though the image is not
congruent to the pre-image it *is* similar to the pre-image. Enlarging a wallet-sized
picture to hang as a portrait on the wall would be an example of a dilation.

Depending on the scale factor, dilations can take small objects and make them larger, but they can also take a larger object and make it smaller. The **scale factor**, also known as the *scalar factor*, measures how much larger or smaller the object will become compared to the pre-image. A dilation changes the distance between points by multiplying the $x$ and $y$ coordinates by the scale factor. You can see this in the following rule for a dilation with a scale factor $k$. The dilations that we will be doing in this chapter are centered at the origin unless otherwise specified.

Rule: $(x,y) \underset{D_k}{\rightarrow} (kx,ky)$

Let's look at a basic example asking us to dilate a point.

**EXAMPLE**

▶ Find the image of $P(5,4)$ under $D_2$.

▶ The scale factor for this dilation is 2. Multiplying the $x$ and $y$ coordinates by 2 will result in the image point $P'(10,8)$.

The next example requires you to find the scale factor of a dilation to help you find an image point.

**EXAMPLE**

▶ A dilation maps $A(2,-3)$ to $A'(10,-15)$. What would be the image of $B(-1,7)$ under the same dilation?

▶ Compare the $x$ and $y$ values from $A$ to $A'$ to find the scale factor.

▶ $A(2,-3) \rightarrow A'(10,-15)$ has a scale factor of 5.

▶ Now apply the scale factor to $B$ to find $B'$.

$B(-1,7) \underset{D_5}{\rightarrow} B'(-5,35)$.

As we said earlier, the scale factor can also make an object smaller. This occurs when the scale factor is a number greater than zero but less than 1. This is shown in the following example.

▶ $\triangle RED$ contains the points $R(2,2)$, $E(6,2)$, and $D(6,8)$. Graph $\triangle RED$ and then graph and state the coordinates of $\triangle R'E'D'$, the image of $\triangle RED$ after a dilation with a scale factor of $\frac{1}{2}$.

$$R(2,2) \underset{D_{\frac{1}{2}}}{\longrightarrow} R'(1,1)$$

$$E(6,2) \underset{D_{\frac{1}{2}}}{\longrightarrow} E'(3,1)$$

$$D(6,8) \underset{D_{\frac{1}{2}}}{\longrightarrow} D'(3,4)$$

**EASY MISTAKE**

Many students think when the scale factor of dilation is a fraction the object will always get larger. That is not true! $\frac{4}{3}$ is a fraction but its value is larger than 1. If we were to dilate something by a scale factor of $\frac{4}{3}$ the object would get larger not smaller.

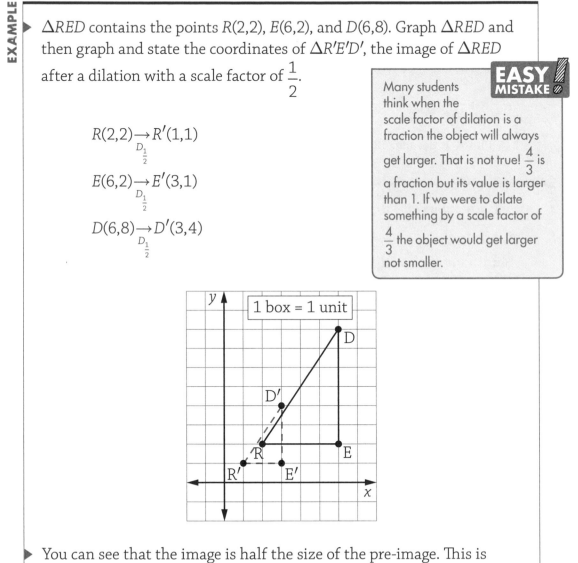

You can see that the image is half the size of the pre-image. This is because the scale factor was one-half.

What do you think happens when the scale factor is a negative number? A common mistake is to think that a negative scale factor means that the image will get smaller. We know that cannot be true because we already learned that a scale factor between 0 and 1 makes the image smaller. The following diagrams compare a dilation with a scale factor of 2 to a dilation with a scale factor of $-2$ applied to $\triangle WIN$.

**EXAMPLE**

$$W(1,2)\underset{D_2}{\rightarrow}W'(2,4)$$

$$I(1,5)\underset{D_2}{\rightarrow}I'(2,10)$$

$$N(5,2)\underset{D_2}{\rightarrow}N'(10,4)$$

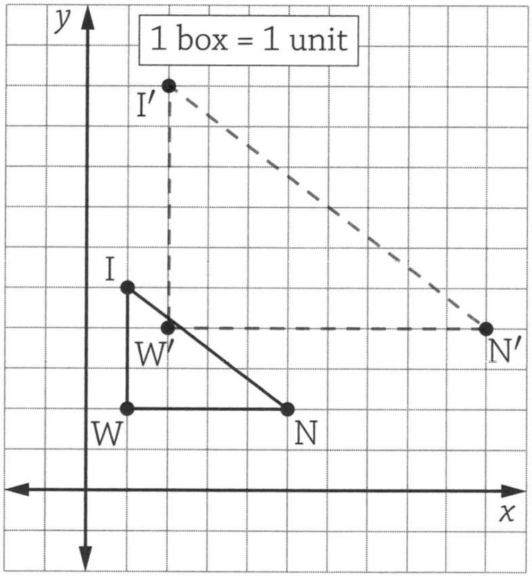

$$W(1,2)\underset{D_{-2}}{\rightarrow}W'(-2,-4)$$

$$I(1,5)\underset{D_{-2}}{\rightarrow}I'(-2,-10)$$

$$N(5,2)\underset{D_{-2}}{\rightarrow}N'(-10,-4)$$

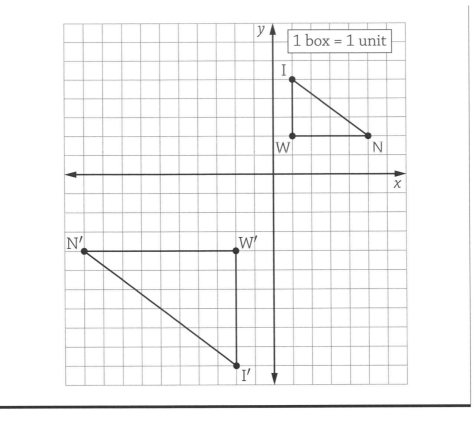

You can see that both dilations in the example result in the image being twice the size of the pre-image. However, when the scale factor is negative, the size of the image is changed *and* the image is rotated 180 degrees.

### Dilations *Not* Centered at the Origin

We previously stated that the dilations performed in this chapter are centered at the origin unless otherwise specified. In this section, we will discuss what happens when the dilation is *not* centered at the origin. There are two methods to performing these dilations. We will go through both methods step by step and then let you decide which one you prefer.

EXAMPLE

The vertices of △GAV are G(2,0), A(1,4), and V(–2,5). Graph △GAV and find the coordinates for △G'A'V', the image of △GAV after a dilation with a scale factor of 2 centered at K(3,2).

**Method 1**

▶ Redraw the axes so that the origin is now at the center of dilation K(3,2).

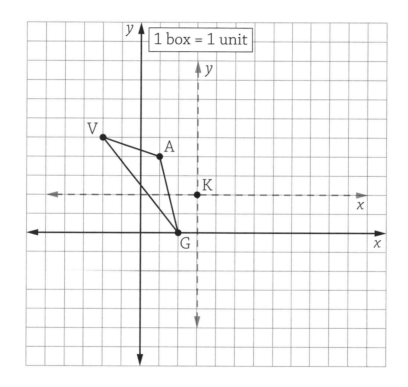

▶ With K(3,2) as the new origin, the points in △GAV are now represented as G(–1,–2), A (–2,2), and V(–5,3).

▶ Now we can dilate $\triangle GAV$ by a scale factor of 2, centered at $K$ (which represents the new origin).

$$G(-1,-2)\underset{D_2}{\rightarrow}(-2,-4)$$

$$A(-2,2)\underset{D_2}{\rightarrow}(-4,4)$$

$$V(-5,3)\underset{D_2}{\rightarrow}(-10,6)$$

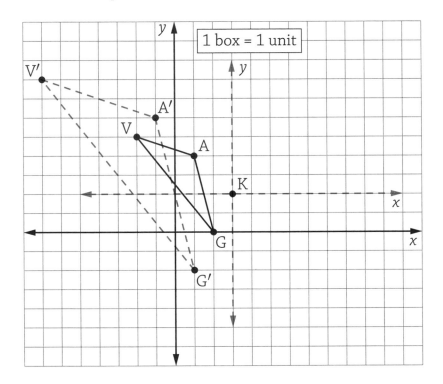

▶ Take the three points that we just found and rename them with the origin in its original location. The final answer is $G'(1,-2)$ $A'(-1,6)$ $V'(-7,8)$.

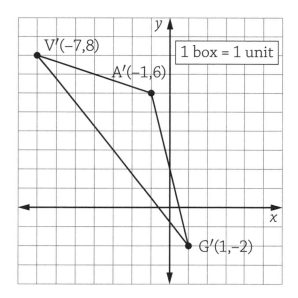

## Method 2

▶ Graph $\triangle GAV$ and $K(3, 2)$.

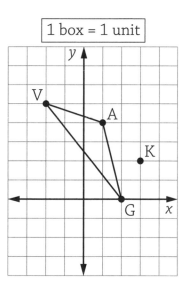

▶ To get from point $K$ to point $G$, you would count one box to the left and two boxes down. From $G$ again, count one box to the left and two boxes down again. Arrive at point $G'(1,-2)$.

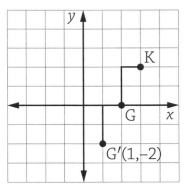

▶ To get from point $K$ to point $A$, you would count two boxes to the left and two boxes up. From $A$ again, count two boxes to the left and two boxes up. Arrive at point $A'(-1,6)$.

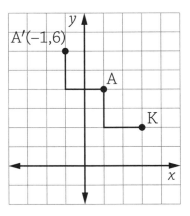

▶ To get from point *K* to point *V*, you would count five boxes to the left and three boxes up. From *V* again, count five boxes to the left and three boxes up. Arrive at point *V'*(−7,8).

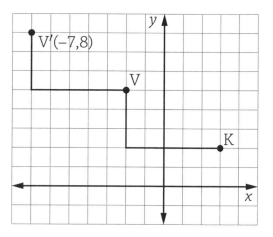

The final answer is *G'*(1,−2) *A'*(−1,6) *V'*(−7,8).

▶ If you noticed in steps two through four, we repeated the pattern from point *K* to each of the pre-images TWO times because the scale factor was 2. If the scale factor was three, we would have repeated each pattern from *K* three times.

## Composition of Transformations

A **composition of transformations** occurs when two or more transformations are performed on the pre-image. For example, we can reflect a triangle over the *x*-axis and then follow it by reflecting the new image over the *y*-axis. This can be written using the notation $r_{y-axis} \circ r_{x-axis} (\triangle WIN)$.

**BTW**

Composition of transformations is not commutative.

The composition is read from right to left. You perform the transformation closest to the point first.

The following table shows △WIN being reflected over the x-axis followed by a reflection over the y-axis. The final column represents the points for △W′I′N′.

| △WIN | $r_{x-axis}$ | $r_{y-axis}$ |
|---|---|---|
| W(1,2) | (1,−2) | W′(−1,−2) |
| I(1,5) | (1,−5) | I′(−1,−5) |
| N(5,2) | (5,−2) | N′(−5,−2) |

Can you name a single transformation that is equivalent to this composition? If you said a rotation of 180 degrees, you are right! If you notice, the image coordinates are negated versions of the pre-image coordinates. We learned earlier in the chapter that when we rotate 180 degrees we negate the x and y coordinates. Therefore, a rotation of 180 degrees centered at the origin is equivalent to reflecting a point through the x-axis and the y-axis.

The following example uses the composition of transformations to locate the image point.

EXAMPLE

▶ The following diagram represents a regular octagon. Lines ℓ and p represent lines of symmetry. Find the image of the given point under each composition.

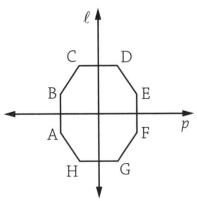

$$R_{90°} \circ r_\ell \ (A)$$
$$r_p \circ R_{225°} \ (E)$$
$$R_{135°} \circ r_\ell \circ R_{-180°}(C)$$

▶ An octagon has eight vertices. To determine the number of degrees it takes to map this octagon onto itself, we must divide 360° by 8. This means that each "turn" or vertex of the octagon is equivalent to a 45° rotation. Don't forget that we always rotate counterclockwise when using a positive degree of rotation.

$R_{90°} \circ r_\ell(A)$ Reading the composition from right to left, we first reflect point $A$ over the line $\ell$. The new image is point $F$. Now we rotate point $F$ 90° counterclockwise. This is an equivalence of rotating two vertices on the octagon. Point $F$ now lands on point $D$. Therefore, the solution is point $D$.

$r_p \circ R_{225°} \ (E)$ Reading the composition from right to left, we first rotate point $E$ 225° counterclockwise—that is, rotating five vertices on the octagon. The image is point $H$. Then we reflect point $H$ over line $p$. The image is point $C$.

$R_{135°} \circ r_\ell \circ R_{-180°} \ (C)$ Reading the composition from right to left, we first rotate point $C$ 180° clockwise (because it is negative)—that is, rotating four vertices on the octagon. The image is point G. Then we reflect point G over line $\ell$, and the new image is point $H$. Finally, we rotate point $H$ 135° counterclockwise. The image is point $E$.

It is also important for us to be able to recognize sequences of transformations that would map two congruent shapes onto one another. Let's look at an example of this.

▶ Describe a sequence of transformations that would map $\triangle ABC$ onto $\triangle A'B'C'$ in the following diagram.

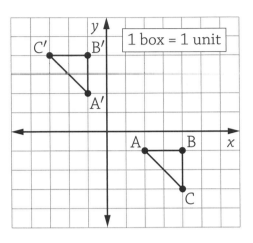

1 box = 1 unit

▶ $\triangle ABC$ is first reflected over the $x$-axis and is then followed by a rotation of 90 degrees counterclockwise. This can be written as a composition: $R_{90°} \circ r_{x\text{-axis}}(\triangle ABC)$.

## EXERCISES

### EXERCISE 10-1

*Find the image of point P(4,25) after each given transformation.*

1. $r_{x-axis}$

2. $r_{y-axis}$

3. $r_{y=x}$

4. $r_{x=5}$

5. $r_{y=20}$

6. $T_{8,-1}$

7. $D_3$

8. $D_{\frac{1}{4}}$

### EXERCISE 10-2

*Use the information provided to find the image of the point.*

1. Given point $R(2,12)$, find $R'$, the image of point $R$ after a rotation of $90°$ around the origin.

2. Given point $S(2,-3)$, find $S'$, the image of point $S$ after a *clockwise* rotation of $90°$ around the origin.

### EXERCISE 10-3

*Indicate whether each of the following statements is true or false.*

1. A dilation is a rigid motion.

2. A translation is a direct isometry.

3. A line reflection is a rigid motion that preserves orientation.

## EXERCISE 10-4

**Answer each of the following questions.**

1. Write the equation for the line of reflection that maps $\triangle ABC$ onto $\triangle A'B'C'$ in the following diagram.

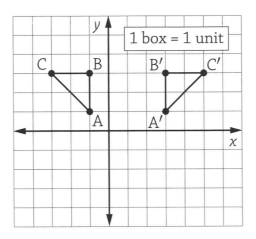

2. The endpoints of $\overline{CD}$ are $C(5,10)$ and $D(12,3)$. Find the endpoints of $\overline{C'D'}$, the image of $\overline{CD}$ after the translation $T_{4,-9}$.

3. Find the image of $G(-15,40)$ under $D_{\frac{1}{5}}$.

4. A dilation maps $A(4,6)$ to $A'(6,9)$. What is the scale factor for the dilation?

5. A dilation maps $M(-2,10)$ to $M'(10,-50)$. Find the image of $B(3,-7)$ under the same dilation.

6. The endpoints of $\overline{HI}$ are $H(3,5)$ and $I(4,1)$. Find the endpoints of $\overline{H'I'}$ under the composition $T_{5,-1} \circ D_2$.

7. Point $P$ is represented by $P(-4,3)$. Find $P'$, the image of $P$ after the composition $T_{1,2} \circ R_{90°} \circ r_{y=x}$.

8. Find the image of $V(3,-3)$ after a reflection over the line $y = -3$. Does this represent an invariant point?

9. Determine a sequence of transformations that maps $\triangle AMZ$ to $\triangle A'M'Z'$ in the following diagram.

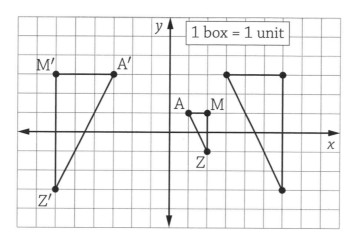

10. Given point $G(3,1)$, find $G'$ after a dilation with a scale factor of 2 centered at $K(4,2)$.

Flashcard App

 # Circle Theorems Involving Angles and Segments

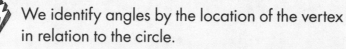 ## MUST ⚡ KNOW

⚡ Circle theorems allow us to measure exterior and interior angles as well as segments of a circle.

⚡ We identify angles by the location of the vertex in relation to the circle.

⚡ A tangent line intersects a circle at exactly one point—the point of tangency. A secant line intersects a circle at two different points.

**L**et's take a break from triangles and quadrilaterals and discuss circles. A **circle** is defined as a set of points equidistant from a fixed point, the fixed point being the center of the circle. This chapter will focus on the different theorems that determine angle measurements and segment lengths formed from segments both inside and outside of the circle.

## Definition of Terms Related to a Circle

We will use the following diagram of circle $O$ to discuss important vocabulary related to circles.

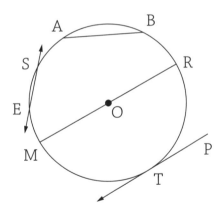

- The segment that joins the center of the circle to a point on the circle is the **radius**.

  $\overline{OR}$ and $\overline{OM}$ are both radii of circle $O$. Radii of the same circle are always congruent.

- A **chord** is a line segment whose endpoints lie on the circle.

  $\overline{AB}$ and $\overline{SE}$ are both chords in circle $O$.

- A **diameter** is a chord that goes through the center of the circle. It is double the length of the radius.

  $\overline{MR}$ is the diameter of this circle.

- An **arc** is part of the **circumference** (outer edge) of the circle.

$\overarc{BR}$ is one of many arcs you can see in our diagram. What can be confusing is that two different arcs can be represented by $\overarc{BR}$. Minor arc $\overarc{BR}$ refers to the smallest distance around the circle from points $B$ to $R$. Major arc $\overarc{BR}$ refers to the largest distance around the circle from points $B$ to $R$. You might be wondering, "How will I know which arc a question is referring to if I don't see the words *major* or *minor*?" We place a third letter in the middle of the two endpoints of the arc to indicate the larger arc. Therefore, we can represent major arc $\overarc{BR}$ as $\overarc{BMR}$.

- A **tangent line** intersects the circle at exactly one point.

  $\overrightarrow{PT}$ is tangent to circle $O$. We call point $T$ the *point of tangency*.

- A **secant line** intersects the circle at two different points.

  $\overleftrightarrow{SE}$ represents a secant line.

A radius is always perpendicular to a tangent line when they meet at the point of tangency.

Let's try an example that uses some of these vocabulary words.

▶ In the accompanying diagram, $\overline{TN}$ is tangent to circle $O$ at point $N$. $ON = 6$ and $AT = 4$. Find the length of $\overline{TN}$.

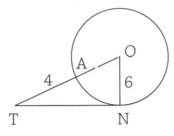

▶ $\overline{ON}$ and $\overline{OA}$ are both radii in circle $O$, and radii of the same circle are congruent. This means $ON = OA = 6$. Radius $\overline{ON}$ intersects tangent $\overline{TN}$ at point $N$; therefore, $\overline{ON} \perp \overline{TN}$. $\angle N$ is a right angle, which means $\triangle TNO$ is a right triangle with hypotenuse $\overline{OT}$. We can find the length of $\overline{TN}$ by using the Pythagorean theorem—but be careful because the problem only gave us the length of $\overline{AT}$. We need the length of hypotenuse $\overline{OT}$ before we can substitute it into the Pythagorean theorem.

$$\overline{OT} = \overline{OA} + \overline{AT}$$

$AT = 4$   (This information was given.)

$OA = 6$

$OT = (6) + (4) = 10$

$$a^2 + b^2 = c^2$$
$$(ON)^2 + (TN)^2 = (OT)^2$$
$$(6)^2 + (TN)^2 = (10)^2$$
$$36 + (TN)^2 = 100$$
$$(TN)^2 = 64$$
$$TN = 8$$

Let's look at an important circle theorem: Two tangent segments drawn from the same external point are equal in length. The next example practices this theorem.

**EXAMPLE**

▶ The following diagram shows △AJK circumscribed about a circle. $\overline{AJ}$, $\overline{JK}$, and $\overline{AK}$ are all tangent to the circle at points B, C, and D. If $JB = 5$, $AB = 7$, and $KD = 4$, what is the perimeter of △AJK?

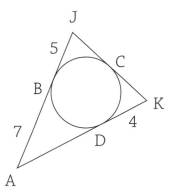

▶ Tangent segments drawn from the same external point are congruent. Therefore,

$$AB = AD = 7$$
$$JB = JC = 5$$
$$KC = KD = 4$$

▶ We can find the perimeter by finding the sum of the sides of △AJK:
$7 + 5 + 7 + 4 + 5 + 4 = 32$.

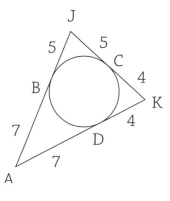

## EXTRA HELP

We can't forget that we will be asked to use our algebra skills within many different types of geometry questions. Let's practice this together using the following example.

▶ In the accompanying diagram, $\overline{JE}$ and $\overline{JT}$ are tangent to the circle at $E$ and $T$. $JE = 10$ and $JT = x^2 - 4x - 2$. Solve for $x$.

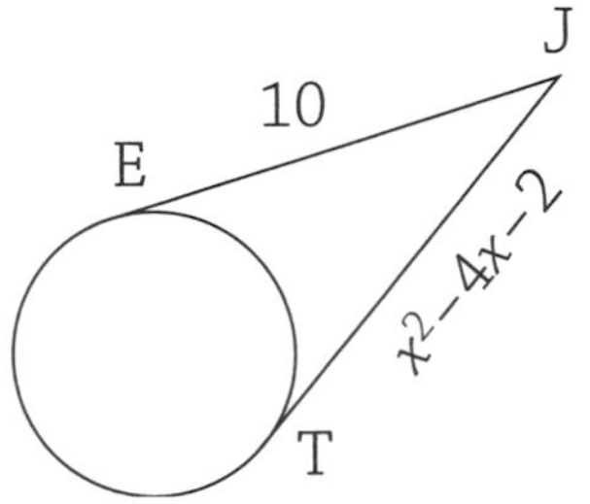

▶ Tangent segments drawn from the same external point are congruent.

$$x^2 - 4x - 2 = 10$$
$$x^2 - 4x - 12 = 0$$
$$(x - 6)(x + 2) = 0$$
$$x = 6 \quad x = -2$$

In the previous examples, we were given information about line segments being tangent to a circle. Our next example requires us to determine if line segments are tangent to a circle.

▶ In the accompanying diagram of circle $A$, $AT = 7$, $TN = 24$, and $AN = 25$. Is $\overline{TN}$ tangent to circle $A$?

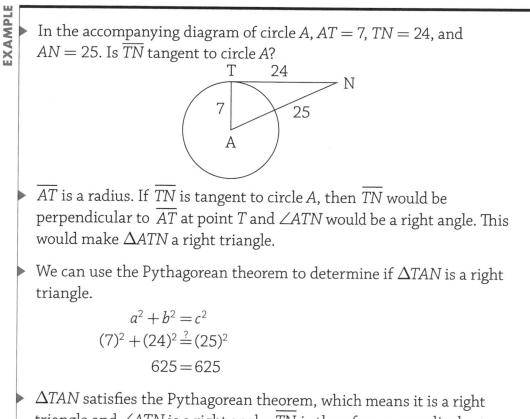

▶ $\overline{AT}$ is a radius. If $\overline{TN}$ is tangent to circle $A$, then $\overline{TN}$ would be perpendicular to $\overline{AT}$ at point $T$ and $\angle ATN$ would be a right angle. This would make $\triangle ATN$ a right triangle.

▶ We can use the Pythagorean theorem to determine if $\triangle TAN$ is a right triangle.

$$a^2 + b^2 = c^2$$
$$(7)^2 + (24)^2 \overset{?}{=} (25)^2$$
$$625 = 625$$

▶ $\triangle TAN$ satisfies the Pythagorean theorem, which means it is a right triangle and $\angle ATN$ is a right angle. $\overline{TN}$ is therefore perpendicular to radius $\overline{AT}$, which means $\overline{TN}$ is tangent to circle $A$.

## Lengths of Intersecting Chords

When two chords intersect in a circle, the product of the segments of one chord is equal to the product of the segments of the other chord. This can be seen in the following diagram of circle $O$ where chords $\overline{CD}$ and $\overline{HR}$ intersect at $E$.

$$CE \times DE = HE \times RE$$

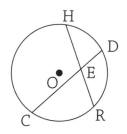

EXAMPLE

In this diagram $CE = 8$, $DE = 3$, and $HE = 6$. Find the length of $\overline{RE}$.

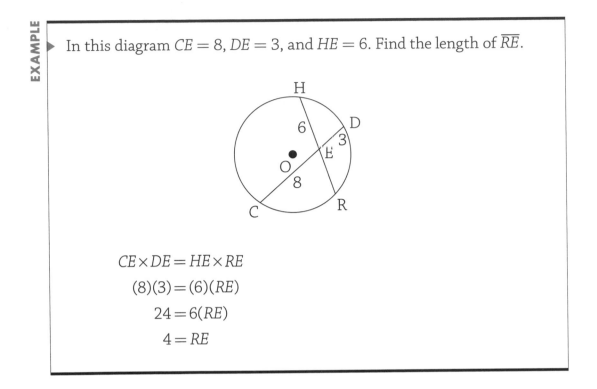

$$CE \times DE = HE \times RE$$
$$(8)(3) = (6)(RE)$$
$$24 = 6(RE)$$
$$4 = RE$$

Let's solve a more challenging question involving this same theorem.

EXAMPLE

In the accompanying diagram of circle $O$, $\overline{CD}$ bisects $\overline{HR}$ at $E$. $HR = 20$ and $\overline{CE}$ is 15 more than $\overline{DE}$. Find the length of $\overline{DE}$.

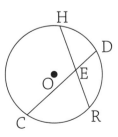

▶ A bisector divides a segment into two congruent segments, which means $HE = RE = 10$.

Let $DE = x$

Let $CE = x + 15$

Quadratic equations must equal 0 before factoring or using the quadratic formula to solve.

**EASY MISTAKE**

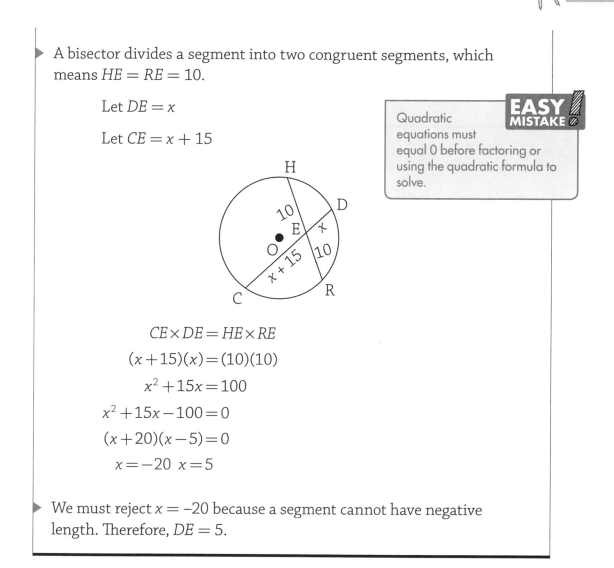

$$CE \times DE = HE \times RE$$
$$(x+15)(x) = (10)(10)$$
$$x^2 + 15x = 100$$
$$x^2 + 15x - 100 = 0$$
$$(x+20)(x-5) = 0$$
$$x = -20 \quad x = 5$$

▶ We must reject $x = -20$ because a segment cannot have negative length. Therefore, $DE = 5$.

## Finding the Length of Secant Segments

When two secant segments are drawn from the same external point, the following rule applies:

(Exterior part of secant 1) × (Entire secant 1) = (Exterior part of secant 2) × (Entire secant 2)

Many students remember this theorem as:

$$(\text{Outside}) \times (\text{Whole}) = (\text{Outside}) \times (\text{Whole})$$

▶ The following diagram of circle $O$ shows secants $\overline{YAN}$ and $\overline{YSK}$.

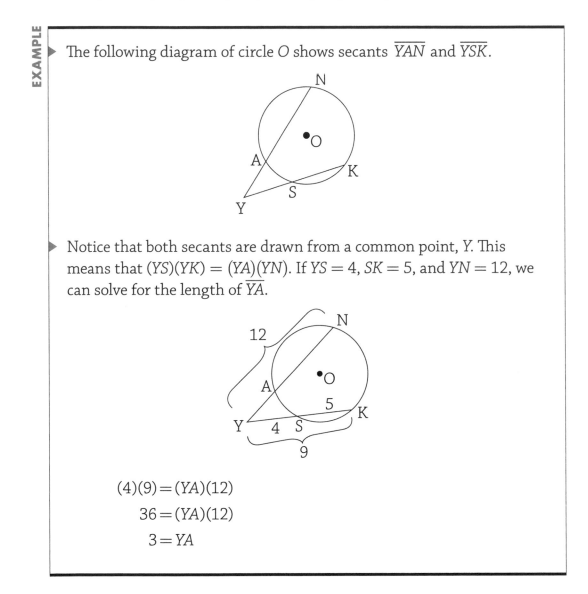

▶ Notice that both secants are drawn from a common point, $Y$. This means that $(YS)(YK) = (YA)(YN)$. If $YS = 4$, $SK = 5$, and $YN = 12$, we can solve for the length of $\overline{YA}$.

$$(4)(9) = (YA)(12)$$
$$36 = (YA)(12)$$
$$3 = YA$$

## EXTRA HELP

Let's solve a more challenging problem involving this same theorem.

> In the accompanying diagram of circle $O$, secants $\overline{YAN}$ and $\overline{YSK}$ are drawn. If $YS = 3$, $SK = 5$, and $AN = 10$, what is the length of $\overline{YA}$?

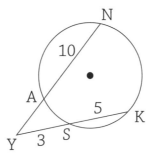

> Remember:
>
> (Exterior part of secant 1) × (Entire secant 1) = (Exterior part of secant 2) × (Entire secant 2)

> Let $x =$ the length of $\overline{YA}$

$$(YS)(YK) = (YA)(YN)$$
$$(3)(8) = (x)(x+10)$$
$$24 = x^2 + 10x$$
$$0 = x^2 + 10x - 24$$
$$(x+12)(x-2) = 0$$
$$x = -12 \quad x = 2$$

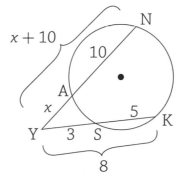

> We must reject $x = -12$ because the length of a segment cannot be negative. Therefore, $YA = 2$.

# Length of Tangent–Secant Segments from an External Point

We can apply the same theorem used for two secants drawn from the same external point to circles that have one tangent and one secant drawn from the same external point. In this case, the exterior length of the tangent is actually the length of the entire tangent. We can, therefore, generate the following rule:

(Tangent) × (Tangent) = (Exterior part of secant) × (Entire secant)

(Tangent)² = (Exterior part of secant) × (Entire secant)

The following diagram shows tangent $\overline{PA}$ and secant $\overline{PBC}$ drawn from external point $P$. This theorem states that $(PA)^2 = (PB)(PC)$.

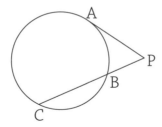

EXAMPLE

If $PC = 9$ and $PB = 4$, we can find the length of $\overline{PA}$ using this theorem:

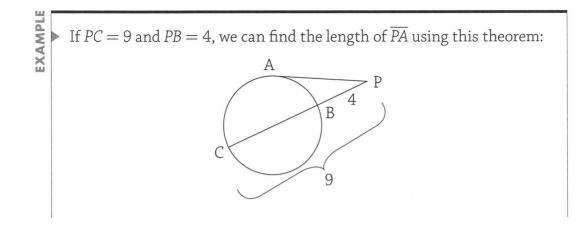

$$(PA)^2 = (PB)(PC)$$
$$(PA)^2 = (4)(9)$$
$$(PA)^2 = 36$$
$$PA = 6$$

# Angles Associated with the Circle

Several theorems can be used to find the measure of different angles and arcs formed from segments both inside and outside a circle. The theorems differ based on the location of the vertex angle. We will begin this section by learning about the angles that are located inside a circle called *central*, *inscribed*, and *vertical*.

### Central Angle

A **central angle** is formed by two radii with the vertex of the angle at the center of the circle.

**EXAMPLE**

In the following diagram, $\angle COD$ is a central angle intercepting $\overset{\frown}{CD}$. When an arc of a circle is intercepted by a central angle, the degree measure of the arc is equal to the degree measure of the central angle. If $m\angle COD = 50°$, then $m\overset{\frown}{CD} = 50°$.

**EASY MISTAKE**

The degree or radian measure of an arc is measuring the arc's curvature not the arc's length.

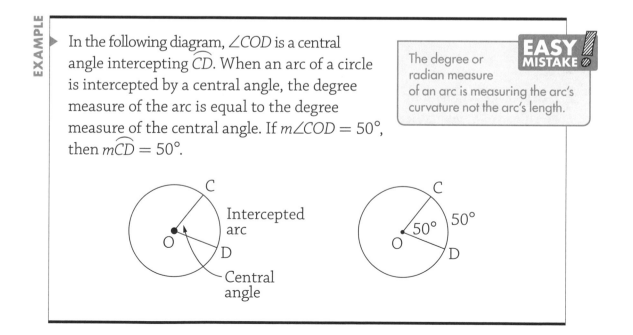

## Inscribed Angle

An **inscribed angle** is an angle formed by two chords in a circle that intersect *on* the circle. This means that the vertex of the angle lies *on* the circle. The degree measure of an inscribed angle is equal to half the degree measure of its intercepted arc.

▶ In the following diagram, $\angle FUN$ is an inscribed angle. If $m\angle FUN = 50°$, then $m\overset{\frown}{FN} = 100°$.

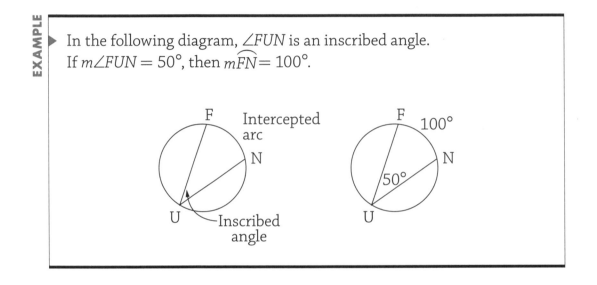

Let's take a look at an example that combines many of the rules we have just learned.

> **BTW**
> *If two chords of a circle are congruent, then their intercepted arcs are congruent.*

## EXTRA HELP

It's important for us to be able to recognize multiple concepts within one circle problem. The following example will do just that.

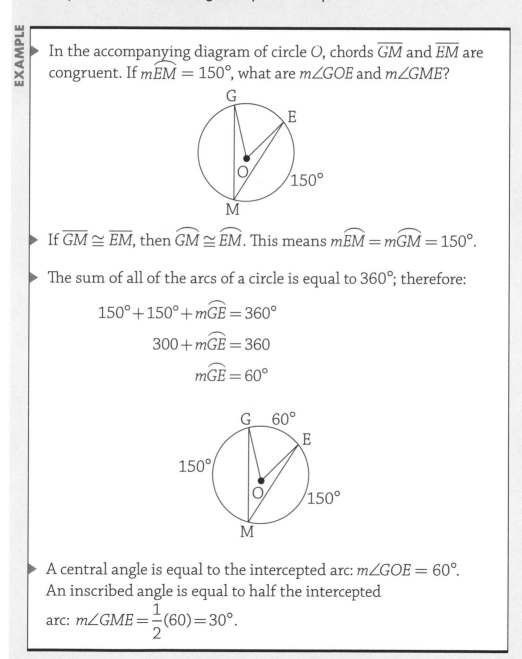

▶ In the accompanying diagram of circle $O$, chords $\overline{GM}$ and $\overline{EM}$ are congruent. If $m\overset{\frown}{EM} = 150°$, what are $m\angle GOE$ and $m\angle GME$?

▶ If $\overline{GM} \cong \overline{EM}$, then $\overset{\frown}{GM} \cong \overset{\frown}{EM}$. This means $m\overset{\frown}{EM} = m\overset{\frown}{GM} = 150°$.

▶ The sum of all of the arcs of a circle is equal to 360°; therefore:

$$150° + 150° + m\overset{\frown}{GE} = 360°$$
$$300 + m\overset{\frown}{GE} = 360$$
$$m\overset{\frown}{GE} = 60°$$

▶ A central angle is equal to the intercepted arc: $m\angle GOE = 60°$. An inscribed angle is equal to half the intercepted arc: $m\angle GME = \dfrac{1}{2}(60) = 30°$.

We learned that the vertex of an inscribed angle always lies on the circle. There is another angle whose vertex also always lies on a circle. It is formed from a tangent meeting a chord, and it follows the same rule as an inscribed angle. The following diagram shows this angle.

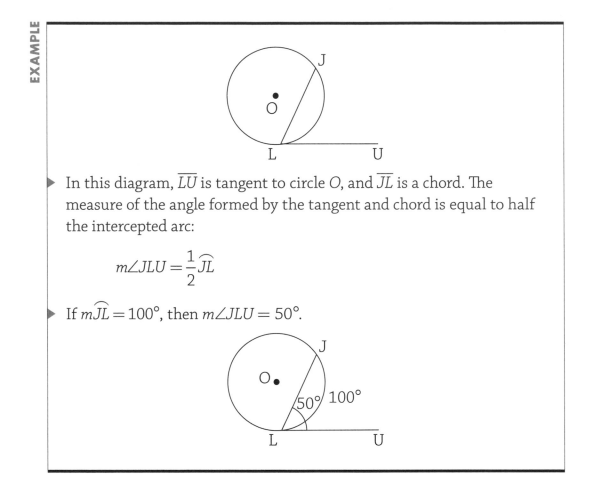

> In this diagram, $\overline{LU}$ is tangent to circle O, and $\overline{JL}$ is a chord. The measure of the angle formed by the tangent and chord is equal to half the intercepted arc:

$$m\angle JLU = \frac{1}{2}\overset{\frown}{JL}$$

> If $m\overset{\frown}{JL} = 100°$, then $m\angle JLU = 50°$.

## Angle Formed by Two Intersecting Chords

When two chords intersect in a circle, they form angles that we like to call **vertical interior angles**.

We already know that vertical angles are congruent. The measure of the congruent angles is equal to half the sum of the measures of the arcs they intercept. In other words, the angle is equal to the average of the intercepted arcs. The accompanying diagram shows chords $\overline{MT}$ and $\overline{AH}$ intersecting in circle O.

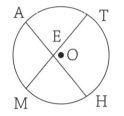

We can see two different pairs of vertical angles: $\angle MEA \cong \angle HET$ and $\angle MEH \cong \angle AET$. If we were looking for the measure of $\angle MEH$, we would set up the following equation:

$$m\angle MEH = \frac{1}{2}(\widehat{AT} + \widehat{MH}) \ \text{ or } \ m\angle MEH = \frac{\widehat{AT} + \widehat{MH}}{2}$$

Let's practice this rule in the following example.

EXAMPLE

▶  In the accompanying diagram of circle O, two chords $\overline{RS}$ and $\overline{TW}$ intersect at E. If $m\widehat{TR} = 80°$ and $m\angle SEW = 120°$, then what is the measure of $\widehat{SW}$?

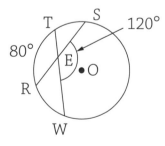

▶ $\angle SEW$ is a vertical angle formed by two intersecting chords.

$$m\angle SEW = \frac{m\overset{\frown}{TR} + m\overset{\frown}{SW}}{2}$$

$$\frac{120}{1} = \frac{80 + m\overset{\frown}{SW}}{2}$$

$$240 = 80 + m\overset{\frown}{SW}$$

$$160° = m\overset{\frown}{SW}$$

## Exterior Angles of a Circle

Angles that are formed outside of a circle—exterior angles—can be formed three different ways: by two secants, a secant and a tangent, or two tangents drawn from the same external point. The same theorem applies to all three scenarios.

The measure of an exterior angle is equal to half the difference of the measures of the intercepted arcs. The difference is always found by subtracting the smaller arc from the larger arc. You can see this same theorem applied to the three different scenarios for an exterior angle in the following diagrams.

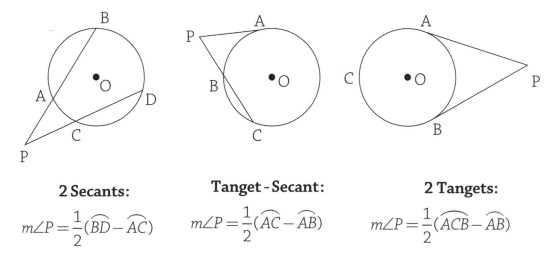

**2 Secants:**

$$m\angle P = \frac{1}{2}(\overset{\frown}{BD} - \overset{\frown}{AC})$$

**Tanget - Secant:**

$$m\angle P = \frac{1}{2}(\overset{\frown}{AC} - \overset{\frown}{AB})$$

**2 Tangets:**

$$m\angle P = \frac{1}{2}(\overset{\frown}{ACB} - \overset{\frown}{AB})$$

Let's find the measure of the exterior angle of a circle in the following example.

▶ In the circle drawn in the accompanying diagram, secants $\overline{ABC}$ and $\overline{ADE}$ intersect at $A$. If $m\widehat{CE} = 120°$ and $m\widehat{BD} = 50°$, what is $m\angle CAE$?

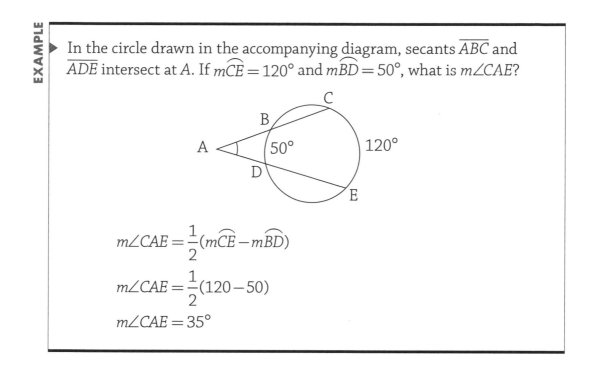

$$m\angle CAE = \frac{1}{2}(m\widehat{CE} - m\widehat{BD})$$

$$m\angle CAE = \frac{1}{2}(120 - 50)$$

$$m\angle CAE = 35°$$

## EXERCISES

### EXERCISE 11-1

*Solve the following problems.*

1. In the accompanying diagram, $\overline{AB}$ is tangent to circle $O$ at $B$. If $AO = 15$ and $AB = 12$, what is the length of the diameter of circle $O$?

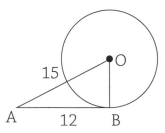

2. The following diagram shows circle $O$ inscribed in $\triangle GTA$ if $SG = 12$, $AI = 8$, and $TN = 15$. Find the perimeter of $\triangle GTA$.

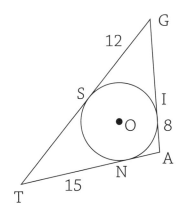

## EXERCISE 11-2

*Use the accompanying diagram and given information to solve the following problems. In circle A, secant* $\overline{SEC}$ *and tangent* $\overline{SN}$ *are drawn.*

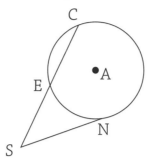

1. $SC = 16$ and $SE = 9$; find $SN$.

2. $SN = 9$ and $SC = 27$; find $SE$.

3. $SE = 2$ and $SN = 4$; find $CE$.

4. $SC = 32$ and $CE = 24$; find $SN$.

## EXERCISE 11-3

*Use the accompanying diagram of circle O to solve the following questions.*

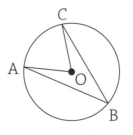

1. $m\angle COA = 75°$; *find* $m\widehat{AC}$.

2. $m\angle COA = 50°$; *find* $m\angle CBA$.

3. $\overline{BA} \cong \overline{BC}$ *and* $m\angle ABC = 40°$; *find* $m\widehat{AB}$.

## EXERCISE 11-4

*Use the accompanying diagram and given information to solve the following problems. In circle O, chords $\overline{LC}$ and $\overline{KU}$ intersect at Y.*

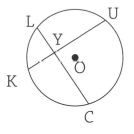

1. $LY = 4$, $CY = 10$, and $KY = 5$. Find the length of $\overline{UY}$.

2. $UY = 6$ and $KY = 4$; $CY$ is 5 more than $LY$. Find the length of $\overline{CY}$.

3. $m\widehat{LU} = 72°$ and $m\widehat{KC} = 88°$; find $m\angle LYU$.

4. $m\widehat{KL} = 47°$ and $m\angle LYK = 82°$; find $m\widehat{UC}$.

5. $m\widehat{CK} = 100°$, $m\widehat{KL} = 60°$, and $m\widehat{LU} = 70°$; find $m\angle CYU$.

## EXERCISE 11-5

*Use the accompanying diagram and given information to solve the following problems. In circle R, secants $\overline{TAC}$ and $\overline{TOD}$ are drawn from external point T.*

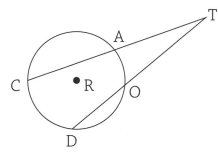

1. $m\widehat{AO} = 56°$ and $m\widehat{CD} = 130°$; find $m\angle T$.

2. If $m\widehat{AO} = 68°$ and $m\angle T = 45°$, what is $m\widehat{CD}$?

3. $DT = 12$, $OT = 8$, and $CT = 16$; find $TA$.

 **Circumference
and the Area
of Circles**

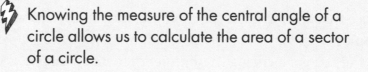 **MUST KNOW**

 The radius—or diameter—is your most useful
tool in calculating the circumference or area of a
circle.

The symbol $\pi$ describes the relationship between
the circumference and diameter of a circle. It is
commonly approximated as 3.14.

Knowing the measure of the central angle of a
circle allows us to calculate the area of a sector
of a circle.

he previous chapter focused on the measure of the angles, arcs, and segments in a circle. This chapter focuses on the area and circumference of a circle. First let's review the definition of a circle and the important vocabulary associated with it.

A circle is a locus of points equidistant from a given point. The given point is the center of the circle. The **radius**, **r**, of a circle is the distance from the center of the circle to any point on the circle. The **diameter**, **d**, of a circle is a segment that passes through the center of the circle, and the endpoints are on the circle. A full revolution of a circle is 360°. A circle is *not* a polygon because it has no sides.

A locus is a set of points that satisfy a given condition.

The **circumference** of a circle is the distance around the circle. To find the distance around a polygon (perimeter), we simply add all the sides. Because there are no sides in a circle, we will calculate the circumference with the following formula:

$$C = 2\pi r \text{ or } C = \pi d$$

The area of a circle is the total space that is enclosed inside the boundaries of the circle and is calculated using the following formula:

$$A = \pi r^2$$

Notice that the formulas for circumference and area both contain $\pi$, a common—and important—constant in mathematics. Spelled out as *pi*, $\pi$ is the ratio of a circle's circumference to its diameter:

$$\pi = \frac{C}{d}$$

Approximated at 3.14, $\pi$ is an irrational number because $\frac{C}{d}$ *cannot* be represented as a fraction with a numerator and denominator that are integers. If we were to try to write it out in decimal form, it would be infinitely long and without any repetition.

Let's practice finding the circumference and area of a circle when we are given the length of the radius.

EXAMPLE

▶ Find, in terms of $\pi$, the circumference and area of a circle that has a radius of 4 inches.

$$C = 2\pi r$$
$$= 2\pi(4)$$
$$= 8\pi \text{ in}$$

$$A = \pi r^2$$
$$= \pi(4)^2$$
$$= 16\pi \text{ in}^2$$

Now let's try a problem where we are not given the length of the radius.

EXAMPLE

▶ If the area of a circle is $49\pi$ ft$^2$, find the length of the diameter.

$$\text{Area} = \pi r^2$$
$$49\pi = \pi(r^2)$$
$$49 = r^2$$
$$7 = r$$

▶ diameter = 2(radius) = 2(7) = 14 feet

## Finding the Area of a Sector

Think of a slice of pizza in a whole pizza pie as the **sector** of a circle. A sector is defined as the part of the circle enclosed by two radii and the intercepted arc and is shown in the following diagram. Because the sector is only part of the whole circle, the area of the sector must be less than that of the whole circle. It will be a fraction of the circle's area. That fraction depends on the degree measure of the central angle for the "slice of pizza."

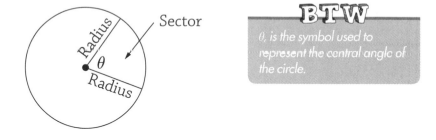

The ratio of the degree measure of the central angle of the sector to the degree measure of the entire circle is directly proportional to the ratio of the area of the sector to the area of the entire circle.

$$\frac{\theta}{360°} = \frac{\text{Area of sector}}{\text{Area of circle}}$$

We can use this proportion to generate a formula for the area of the sector of a circle:

$$\frac{\theta}{360°} = \frac{\text{Area of sector}}{\pi r^2}$$

$$\text{Area of sector} \frac{(\cancel{360°})}{\cancel{360°}} = \frac{\theta(\pi r^2)}{360°}$$

Area of a sector:     $A = \dfrac{\theta}{360°}\left(\pi r^2\right)$

Let's find the area of the sector of a circle of a circle given the central angle and radius of the circle.

EXAMPLE

▶ Find the area of the sector of a circle that has a central angle of 60° and radius of 12 centimeters (cm). Leave your answer in terms of $\pi$.

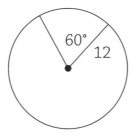

▶ We will solve this in two different ways. First let's solve this by setting up a proportion:

$$\frac{60}{360} = \frac{\text{Area of sector}}{\pi(12)^2}$$

$$\frac{1}{6} = \frac{\text{Area of sector}}{144\pi} \; ; \text{multiple each side by } 144\pi$$

$$\frac{144\pi}{6} = \text{Area of sector}$$
$$24\pi \text{ cm}^2 = \text{Area of sector}$$

▶ We can also solve the problem using the formula for the area of the sector of a circle:

$$\text{Area of a sector} = \frac{60}{360}(\pi(12)^2)$$
$$\text{Area of a sector} = \frac{1}{6}(144\pi)$$
$$\text{Area of a sector} = 24\pi \text{ cm}^2$$

Now let's use the formula to find the radius of a circle given the central angle and area of a sector of the circle.

EXAMPLE

A sector of a circle has a central angle that measures 40° and an area of $1.96\pi$ in². What is the radius of the circle?

$$\text{Area of sector} = \frac{\theta}{360}(\pi r^2)$$

$$1.96\pi = \frac{40}{360}(\pi r^2)$$

$$\left(\frac{9}{1}\right)1.96\pi = \frac{1}{\cancel{9}}(\pi r^2)\left(\frac{\cancel{9}}{1}\right)$$

$$\frac{17.64\cancel{\pi}}{\cancel{\pi}} = \frac{\cancel{\pi}r^2}{\cancel{\pi}}$$

$$\sqrt{17.64} = \sqrt{r^2}$$

$$4.2 \text{ in} = r$$

**EASY MISTAKE**

There are times when we want to divide by multiple values when isolating a variable. Either divide them in separate steps or make sure to put them in parentheses.

## EXTRA HELP

This next example shows how to find the angle of a sector of a circle when you are given the area of the sector and the radius of the circle.

EXAMPLE

What is the measure of the angle of a sector of a circle if the area of the sector is $64\pi$ m² and the radius of the circle is 10 meters?

$$\text{Area of a sector} = \frac{\theta}{360}(\pi r^2)$$

$$64\pi = \frac{\theta}{360}(\pi(10)^2)$$

$$\left(\frac{360}{1}\right)64\pi = \frac{\theta}{\cancel{360}}(100\pi)\left(\frac{\cancel{360}}{1}\right)$$

$$\frac{23,040\cancel{\pi}}{100\cancel{\pi}} = \frac{\theta(10\cancel{0}\pi)}{\cancel{100}\pi}$$

$$230.4° = \theta$$

## Finding the Length of the Arc of a Sector

As we previously learned, a sector is only part of the whole circle. The length of the arc of a sector, sometimes called the **length of the arc intercepted by the central angle of a circle**, must be less than that of the circumference of the circle. It will be a fraction of the circle's circumference. Just like the area of the sector, the length of the arc of a sector also depends on the degree measure of the central angle.

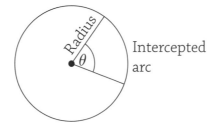

The ratio of the degree measure of the central angle of the sector to the degree measure of the entire circle is directly proportional to the ratio of the length of the sector to the circumference of the circle.

$$\frac{\theta}{360°} = \frac{\text{arc length}}{\text{circumference}}$$

$$\frac{\theta}{360°} = \frac{\text{arc length}}{2\pi r}$$

We can use this proportion to generate a formula to find the length of the arc of a sector of a circle:

$$\frac{\theta}{360°} = \frac{\text{arc length}}{2\pi r}$$

$$\frac{\text{arc length }(\cancel{360°})}{\cancel{360°}} = \frac{\theta(2\pi r)}{360°}$$

$$\text{arc length} = \frac{\theta}{360°}(2\pi r)$$

Let's find the length of the arc of a sector of a circle given the central angle and radius of the circle.

▶ Find, to the nearest tenth of a cm, the length of the arc of the sector of a circle that has a central angle of 30° and radius of 10 cm.

▶ We will solve this two ways. First let's solve this by setting up a proportion:

$$\frac{\theta}{360°} = \frac{\text{arc length}}{2\pi r}$$

$$\frac{30}{360} = \frac{\text{arc length}}{2\pi(10)}$$

$$\frac{1}{12} = \frac{\text{arc length}}{20\pi}$$

$$\frac{20\pi}{12} = \text{arc length}$$

$$5.2 \text{ cm} = \text{arc length}$$

**EASY MISTAKE**

Once a proportion is set up, cross multiplying will help you solve for the missing piece.

▶ Now let's solve the problem using the formula for the length of the arc of a sector of a circle:

$$\text{Arc length} = \frac{\theta}{360}(2\pi r)$$

$$\text{Arc length} = \frac{30}{360}(2\pi)(10)$$

$$\text{Arc length} = 5.2 \text{ cm}$$

The formula will also help us find the radius of a circle given the central angle and the length of the arc of a sector of a circle.

▶ The central angle of a circle is 100° and intercepts an arc that measures $10\pi$ inches. What is the measure of the radius of the circle?

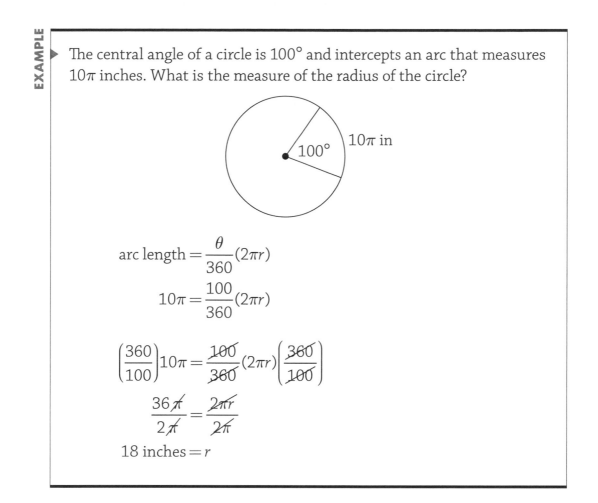

$$\text{arc length} = \frac{\theta}{360}(2\pi r)$$

$$10\pi = \frac{100}{360}(2\pi r)$$

$$\left(\frac{360}{100}\right)10\pi = \frac{\cancel{100}}{\cancel{360}}(2\pi r)\left(\frac{\cancel{360}}{\cancel{100}}\right)$$

$$\frac{36\cancel{\pi}}{2\cancel{\pi}} = \frac{2\cancel{\pi} r}{2\cancel{\pi}}$$

$$18 \text{ inches} = r$$

The **perimeter of a sector** is the sum of the lengths of the three pieces that enclose the sector of the circle. This means that the perimeter of a sector of a circle equals the sum of two radii of the circle and the length of the arc of the sector.

Find the perimeter of a sector of a circle that has a central angle of 40° and a radius of 8 meters. Round the perimeter to the nearest tenth of a meter.

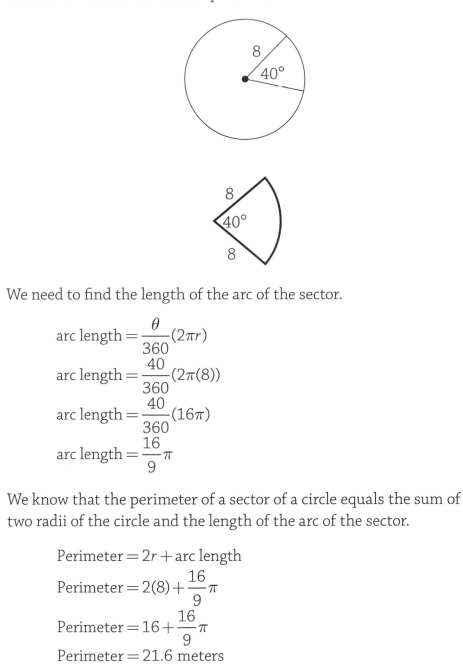

▶ We need to find the length of the arc of the sector.

$$\text{arc length} = \frac{\theta}{360}(2\pi r)$$

$$\text{arc length} = \frac{40}{360}(2\pi(8))$$

$$\text{arc length} = \frac{40}{360}(16\pi)$$

$$\text{arc length} = \frac{16}{9}\pi$$

▶ We know that the perimeter of a sector of a circle equals the sum of two radii of the circle and the length of the arc of the sector.

$$\text{Perimeter} = 2r + \text{arc length}$$

$$\text{Perimeter} = 2(8) + \frac{16}{9}\pi$$

$$\text{Perimeter} = 16 + \frac{16}{9}\pi$$

$$\text{Perimeter} = 21.6 \text{ meters}$$

# Standard Form of a Circle

The equation of a circle can be written in two forms—standard and general. Let's first learn the **standard form** for the equation of a circle, also known as the **center-radius** form of a circle.

$$(x-h)^2 + (y-k)^2 = r^2$$

The center of the circle is represented by $(h,k)$ and the radius by $r$. When a circle is written in this form, the center and radius of the circle can be easily determined. We can see this in the following example.

**EXAMPLE**

▶ Find the center and radius of a circle whose equation is $(x - 2)^2 + (y + 3)^2 = 25$.

▶ If we compare our equation to the standard form of a circle, we can find the center of the circle by solving for $h$ and $k$.

$$\left(x\boxed{-h}\right)^2 + (y\boxed{-k})^2 = r^2$$
$$\left(x\boxed{-2}\right)^2 + (y\boxed{+3})^2 = 25$$
$$\frac{-h}{-1} = \frac{-2}{-1} \qquad \frac{-k}{-1} = \frac{3}{-1}$$
$$h = 2 \qquad k = -3$$
$$\text{Center } (2, -3)$$

▶ Do you notice that the center of the circle ends up having the opposite signs of the numbers in the parenthesis of the equation of the circle? Many students use this trick to quickly find the center of the circle.

▶ Now let's find the radius of the circle. Remember: We are comparing the equation of our circle to the standard form of a circle.

$$(x-h)^2 + (y-k)^2 = \boxed{r^2}$$
$$(x-2)^2 + (y+3)^2 = \boxed{25}$$
$$\sqrt{r^2} = \sqrt{25}$$
$$r = 5$$

> **EASY MISTAKE**
>
> Cutting a value in half is not the same as square rooting the value. When finding the radius from the equation of a circle we square root the value. We cut the value in half when we are given the diameter of a circle and wish to find its radius.

Our next example shows how to write the equation of a circle given the center and radius.

**EXAMPLE**

▶ Write the equation of a circle given that the center is (1,0) and the radius is 7.

$$(x - h)^2 + (y - k)^2 = r^2$$
$$(x - 1)^2 + (y - 0)^2 = 7^2$$
$$(x - 1)^2 + y^2 = 49$$

> **BTW**
>
> The radius of a circle can be rational or irrational. If your radius is irrational, you may be asked to express it in simplest radical form.

## General Form of a Circle

The **general form** of a circle is $x^2 + y^2 + Dx + Ey + F = 0$, where $D$, $E$, and $F$ are constants. Do you notice that there are no parentheses in the general form of a circle? That is because the general form of a circle is obtained by distributing the parentheses in the standard form of a circle and getting them equal to 0. This means that we can convert from one form of a circle to another.

Let's convert an equation of a circle in standard form into general form.

EXAMPLE

▶ Write the general form of a circle whose equation is $(x - 3)^2 + (y + 5)^2 = 10$.

▶ Remember: All we have to do to get this into general form is to get rid of the parentheses through multiplication and then get the equation equal to 0.

$$(x-3)^2 + (y+5)^2 = 10$$
$$(x-3)(x-3) + (y+5)(y+5) = 10$$
$$x^2 - 6x + 9 + y^2 + 10y + 25 = 10$$
$$x^2 + y^2 - 6x + 10y + 34 = \cancel{10}$$
$$\underline{-10 \quad -\cancel{10}}$$
$$x^2 + y^2 - 6x + 10y + 24 = 0$$

When we are looking for the center and radius of a circle, we want the circle written in standard form. It is slightly more challenging to convert from the general form of a circle to the standard form of a circle. Can you think about a way to go from an equation without parentheses to an equation with parentheses squared? Did you guess **completing the square**? If you did, you are correct! Let's try this in the following example.

EXAMPLE

▶ What is the center and radius of a circle whose equation is $x^2 + y^2 - 8x + 4y - 16 = 0$?

▶ To find the center and radius of the circle, we must write the equation in standard form. We will do this by completing the square. We will have to complete the square for the $x$s and the $y$s in the equation, so we will rearrange the equation to group the variables together. Remember that the standard form of a circle does not equal zero, so we will add the 16 to the other side:

$$(x^2 - 8x) + (y^2 + 4y) = 16$$

▶ When we complete the square, we are adding a number to the expression to make it a perfect square binomial. To find this number, we divide the coefficient of the variable by 2 and square it. Let's first try this with the $x$s: The coefficient of $x$ is $-8$; $\dfrac{-8}{2} = -4$ and $(-4)^2 = 16$. This means we will be adding 16 to the $x$ expression. It will now be $x^2 - 8x + 16$. To keep the equation balanced, we must also add 16 to the other side of the equation.

▶ Now let's try this with the $y$s: The coefficient of $y$ is 4; $\dfrac{4}{2} = 2$ and $(2)^2 = 4$. This means we will be adding 4 to the $y$ expression. It will now be $y^2 + 4y + 4$. To keep the equation balanced, we must also add 4 to the other side of the equation.

▶ All together, completing the square makes the equation look like:

$$(x^2 - 8x + \mathbf{16}) + (y^2 + 4y + \mathbf{4}) = 16 + \mathbf{16} + \mathbf{4}$$

▶ Now we can factor our equation:

$$(x - 4)(x - 4) + (y + 2)(y + 2) = 36$$

▶ This can now be written in standard (center-radius) form:

$$(x - 4)^2 + (y + 2)^2 = 36$$

▶ In this form, we can find the center and the radius.

Center: $(4, -2)$

$r^2 = 36$

radius $= 6$

## Graphing a Circle on the Coordinate Plane

When the equation of a circle is already in center-radius form, we can sketch a graph of the circle using its center and radius. We know the length from the center of a circle to any point on the circle

**BTW**

*Although you only need one point that lands on the circle and the center of the circle to draw the circle, you can easily graph four points on the circle: one above the center, one below the center, one to the left of the center, and one to the right of the center.*

is defined as the radius of the circle. Therefore, to graph a circle, we can plot the center of the circle and a point a radius length away from the center and draw the circle using a compass.

Let's graph a circle using the following example.

▶ Graph the circle $(x + 2)^2 + (y - 6)^2 = 16$.

▶ This equation is written in standard form, which means we can easily find the center and radius. The center is $(-2,6)$, and the radius is $\sqrt{16} = 4$. Graph the center by plotting the point $(-2,6)$ on graph paper. Because the length of the radius is 4, count four units to the left, right, up, and down from the center. Sketch the graph of the circle by connecting the four points with a compass.

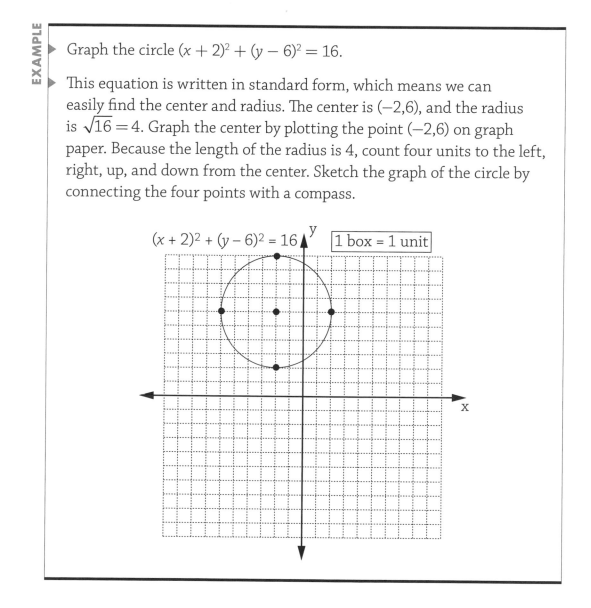

$(x + 2)^2 + (y - 6)^2 = 16$

1 box = 1 unit

We can use the equation of a circle to determine if a point would land inside the circle (interior point), outside the circle (exterior point), or on the circle. It is all based on the radius of the circle. Plug the point into the $x$ and $y$ values of the equation of the circle and compare it to the $r^2$.

The following diagram shows how we can use the equation of a circle to determine where a point is located.

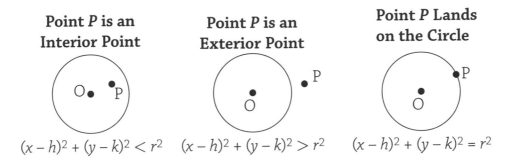

**Point $P$ is an Interior Point**

**Point $P$ is an Exterior Point**

**Point $P$ Lands on the Circle**

$(x - h)^2 + (y - k)^2 < r^2 \qquad (x - h)^2 + (y - k)^2 > r^2 \qquad (x - h)^2 + (y - k)^2 = r^2$

Let's determine where points are located in relation to the given equation of a circle. We will be able to decide if a point is an exterior point, an interior point, or a point that lands on the circle. Our first example will look at a point that lands outside a circle.

EXAMPLE

Show that the point $(-7,1)$ is an exterior point for a circle whose equation is $(x + 1)^2 + (y + 7)^2 = 97$.

By plugging the $x$ and $y$ values of the point into the equation of the circle, we can see that the left side of the equation is greater than the right side of the equation. This means $(-7,1)$ is an exterior point to the circle because it lies outside the circle.

$$((-7)+1)^2 + ((1)+7)^2 > 97$$
$$36 + 64 > 97$$
$$100 > 97$$

Now let's look at a point that lands inside a circle.

▶ Show that the point (4,−2) lands on the circle whose equation is $(x + 1)^2 + (y + 7)^2 = 97$.

▶ By plugging the $x$ and $y$ values of the point into the equation of the circle, we can see that the left side of the equation is less than the right side of the equation. This means (4,−2) is an interior point to the circle because it lies inside the circle.

$$((4)+1)^2 + ((-2)+7)^2 < 97$$
$$(5)^2 + (5)^2 < 97$$
$$50 < 97$$

Lastly, let's look at a point that lands on a circle.

▶ Show that the point (3,2) is an interior point for a circle whose equation is $(x + 1)^2 + (y + 7)^2 = 97$.

▶ By plugging the $x$ and $y$ values of the point into the equation of the circle, we can see that the left side of the equation equals the right side of the equation. This means (3,2) lands on the given circle.

$$((3)+1)^2 + ((2)+7)^2 = 97$$
$$(4)^2 + (9)^2 = 97$$
$$97 = 97$$

## EXERCISES

### EXERCISE 12-1

**Solve the following problems.**

1. Circle $O$ has a radius of 14. Find the area of the circle to the nearest hundredth.

2. Find the circumference of a circle whose diameter is 5 to the nearest tenth.

3. The area of a circle is $81\pi$ ft$^2$; find the diameter of the circle.

### EXERCISE 12-2

**Use the following diagram of a circle with a radius of 16 and a central angle of 100° to solve these problems.**

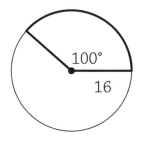

1. Find the area of the sector in terms of $\pi$.

2. Find the arc length of the sector in terms of $\pi$.

3. Find the perimeter of the sector in terms of $\pi$.

## EXERCISE 12-3

*Solve the following problems based off a circle whose equation is*
$(x - 5)^2 + (y + 10)^2 = 20.$

1. Is the equation of this circle in standard or general form?

2. What is the center of this circle?

3. What is the radius of this circle in simplest radical form?

4. Rewrite the circle in general form.

## EXERCISE 12-4

*Follow the instructions for each problem.*

1. Write the equation of a circle in standard form whose center is the origin and radius is 1.

2. Write the equation of a circle in standard form whose center is $(0, \frac{4}{3})$ and radius is $\sqrt{17}$.

3. Write the equation of a circle in standard form whose center is $(-2, -5)$ and diameter is 6.

4. Sketch the graph of $(x + 3)^2 + (y + 1)^2 = 16$.

## EXERCISE 12-5

*For these problems, determine whether each point is in the interior of the circle, exterior of the circle, or on the circle whose equation is* $(x + 4)^2 + (y - 1)^2 = 20.$

1. $(-2, 5)$

2. $(0, 8)$

3. $(-5, 3)$

## EXERCISE 12-6

### Solve the following problems.

1. What is the center and radius of a circle whose equation is
   $x^2 + y^2 - 4x + 12y - 9 = 0$?

2. What is the degree measure of the angle of a sector of a circle whose area
   is $144\pi$ cm$^2$ and radius is 18 cm?

3. A sector of a circle has an area of $400\pi$ m$^2$. The angle of the sector is 45°.
   What is the radius of the circle rounded to the nearest tenth of a meter?

Flashcard
App

# Volume of Three-Dimensional Shapes

## MUST ⚡ KNOW

⚡ Volume formulas allow us to determine the space inside many real-life, three-dimensional (3D) objects.

⚡ We can turn two-dimensional (2D) objects into 3D figures by rotating them around a line.

⚡ The volume of cylinders as well as rectangular and triangular prisms can be calculated using the formula $V = Bh$, where $B$ represents the area of the base and $h$ represents the height.

I n this chapter, we will discuss some of the more popular 3D figures and the formulas needed to calculate their volumes. Cones, cylinders, spheres, and prisms are all three-dimensional objects that are seen in everyday life. An ice cream cone, a tissue box, a basketball, or a can of soup are all real-world examples of the 3D shapes we will be learning about in this chapter.

## Cones

Have you ever wondered how much more ice cream a big waffle cone holds as compared to a regular-sized cone? The only way to answer this question is to find the volume of the different-sized cones. Volume tells us how much space is inside a 3D figure. All the cones we will discuss in this chapter will be right circular cones as opposed to oblique cones. A right circular cone has a height that is perpendicular to the center of the circular base and can be seen in the following diagram.

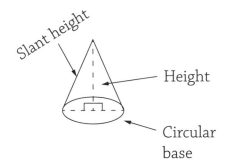

Volume of a right circular cone: $V = \dfrac{1}{3}\pi r^2 h$

Let's use the formula for the volume of a right circular cone to find the volume of a cone given the height and radius of the cone.

The local ice cream shop sells ice cream cones that have a height of 4 inches and a radius of 2 inches. How much ice cream can fit inside the cone? Round your answer to the nearest hundredth of a cubic inch.

▶ We always find it helpful to draw a diagram to help solve the problem.

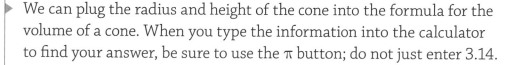

▶ We can plug the radius and height of the cone into the formula for the volume of a cone. When you type the information into the calculator to find your answer, be sure to use the π button; do not just enter 3.14.

$$V = \frac{1}{3}\pi r^2 h$$

$$\frac{1}{3}\pi(2)^2(4) = 16.76 \text{ cubic inches}$$

## EXTRA HELP

It's easy to confuse the height of a cone with the slant height of a cone. The following example should clear up any confusion.

**EXAMPLE**

▶ Find the slant height of a cone that has a radius of 6 feet and a height of 8 feet.

▶ The diagram at the beginning of the chapter shows that the slant height of a right circular cone is measured along the face of the cone and can be described as the distance from the top (apex) of the cone to the base of the cone.

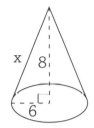

▶ Notice that the height of the cone is perpendicular to the base of the cone and forms a right triangle with the slant height being the hypotenuse. We can therefore use the Pythagorean theorem to solve for the slant height of this right circular cone.

$$\text{leg}^2 + \text{leg}^2 = \text{hypotenuse}^2$$
$$(6)^2 + (8)^2 = x^2$$
$$36 + 64 = x^2$$
$$\sqrt{100} = \sqrt{x^2}$$
$$10 = x$$

▶ The slant height of the cone is 10 feet.

## Cylinders

Let's look at a different 3D figure, the **cylinder**. Picture a can of soda, a can of soup, or the tube inside a paper towel roll. These are all examples of cylinders. They all represent solid shapes with two parallel circular bases.

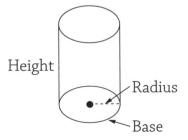

In this section, we will learn how to calculate the surface area and volume of a cylinder. We already know that volume is the amount of space inside the figure. The surface area is the area of all 2D figures that create the 3D figure. You can think of surface area as the amount of wrapping paper needed to wrap the figure.

If we were to unravel a cylinder, we would see that a cylinder is really a rectangle that was rolled together until the opposite sides met. Therefore, to find the surface area of a cylinder, we really need to find the sum of the area of a rectangle whose base is the circumference of the circle and whose height is the height of the cylinder, and the area of the two circles at the top and bottom of the cylinder.

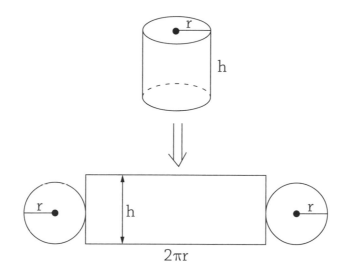

Surface area of cylinder: $S = 2\pi rh + 2\pi r^2$

The following example uses the formula to find the surface area of a cylinder given the radius and height of the cylinder.

**BTW**

*The label on a water bottle is wrapped around the bottle in a cylindrical shape. If you rip off the label and lay it flat on a table, it is a rectangle.*

▶ Find the surface area ($S$) of a right circular cylinder that has a radius of 18 cm and a height of 12 cm. Round your answer to the nearest tenth of a cm.

▶ We will draw a picture of the cylinder to help us solve the problem.

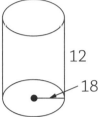

Surface area of a cylinder = Area of the rectangle
　　　　　　　　　　　　　 + 2 (Area of the circular base)

$S = 2\pi rh + 2\pi r^2$

$= 2\pi(18)(12) + 2\pi(18)^2$

$= 3{,}392.9$ cubic cm$^2$

Calculating the volume ($V$) of a cylinder will help us determine how much water is needed to fill a cylindrical pool or the amount of candy that can fit in a cylindrical jar. The volume of a cylinder is found by multiplying the area of the circular base by the height of the cylinder. Sometimes we see the formula for the volume of a cylinder written as $V = Bh$. In this formula, the capital $B$ represents the area of the base of the cylinder, and the $h$ represents the height of the cylinder. Because the base of a cylinder is a circle and the area of a circle is $\pi r^2$, the formula can be written as:

$$V = \pi r^2 h$$

Let's apply this formula in the following example.

EXAMPLE

▶ The neighbor's swimming pool is circular with a diameter of 30 feet. The pool is 6 feet deep throughout the entire pool. How many cubic feet of water are needed to fill the pool all the way to the top? Leave your answer in terms of π.

▶ Let's draw a diagram of the swimming pool.

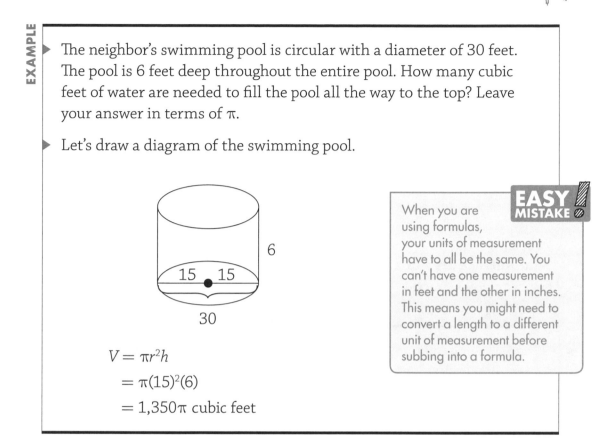

**EASY MISTAKE**

When you are using formulas, your units of measurement have to all be the same. You can't have one measurement in feet and the other in inches. This means you might need to convert a length to a different unit of measurement before subbing into a formula.

$$V = \pi r^2 h$$
$$= \pi (15)^2 (6)$$
$$= 1{,}350\pi \text{ cubic feet}$$

## Prisms

The next 3D figures we will be discussing are prisms. There are two types of prisms we will learn about. The first is a **rectangular prism**. This means that the parallel bases of the prism are rectangles, and the four sides of the prism are rectangles as well.

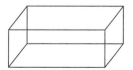

The second type of prism is a **triangular prism**. This means that the parallel bases of the prism are triangles, and the three sides of the prism are rectangles.

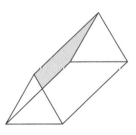

The **surface area of a prism** is found by adding the areas of each face of the prism. Let's practice this using the following example.

▶ The following diagram shows a rectangular prism 10 inches high with a rectangular base 20 inches by 8 inches. Find the surface area of the prism.

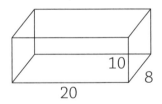

▶ The surface area of the prism will be calculated by finding the sum of the areas of each face of the prism. The two bases of the prism are rectangles measuring 20 inches by 8 inches. Two sides of the prism measure 20 inches by 10 inches, and the other two sides of the prism measure 8 inches by 10 inches. The area of a rectangle is $A = bh$. Therefore, the surface area of the prism is:

$$S = 2(20)(8) + 2(20)(10) + 2(8)(10)$$

▶ The surface area equals 880 square inches.

Let's apply this same concept of surface area to triangular prisms.

Find the surface area of the accompanying triangular prism.

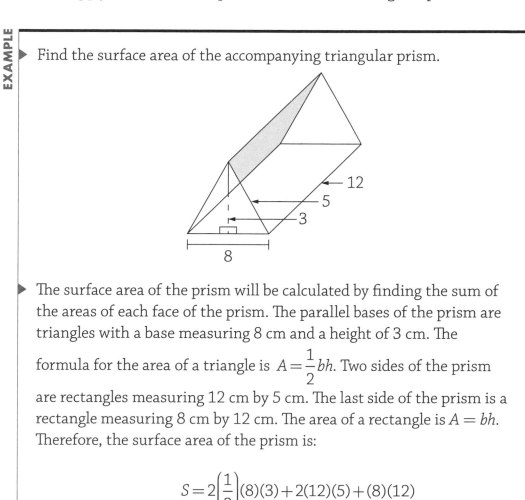

The surface area of the prism will be calculated by finding the sum of the areas of each face of the prism. The parallel bases of the prism are triangles with a base measuring 8 cm and a height of 3 cm. The formula for the area of a triangle is $A = \frac{1}{2}bh$. Two sides of the prism are rectangles measuring 12 cm by 5 cm. The last side of the prism is a rectangle measuring 8 cm by 12 cm. The area of a rectangle is $A = bh$. Therefore, the surface area of the prism is:

$$S = 2\left(\frac{1}{2}\right)(8)(3) + 2(12)(5) + (8)(12)$$

The surface area equals 240 square centimeters.

Now that we understand how to calculate the surface area of rectangular and triangular prisms, let's learn how to calculate the volume of rectangular and triangular prisms. The formula for the volume of a prism is $V = Bh$, where $B$ is the area of the base and $h$ is the height of the prism. We can therefore generate the following formulas for the volumes of prisms:

Volume of a rectangular prism: $V = l \cdot w \cdot h$

Volume of a triangular prism:

$$V = \frac{1}{2} \text{ (Base of triangle) (Height of triangle) (Height of prism)}$$

It is time for us to practice calculating the volume of rectangular and triangular prisms. Our first example calculates the volume of a rectangular prism.

▶ A rectangular prism has a square base whose length is 8 feet. If the height of the prism is 10 feet, what is the volume of the prism?

▶ Because the base is a square, we know all the sides of the base are equal in length. Let's draw a picture to help us solve this problem.

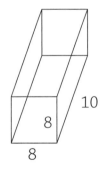

$V = Bh$

$V = l \cdot w \cdot h$

$V = 8(8)(10)$

$V = 640$ cubic feet

Let's now try to find the volume of a triangular prism.

**EXAMPLE**

▶ Find the volume of the prism shown in the accompanying diagram.

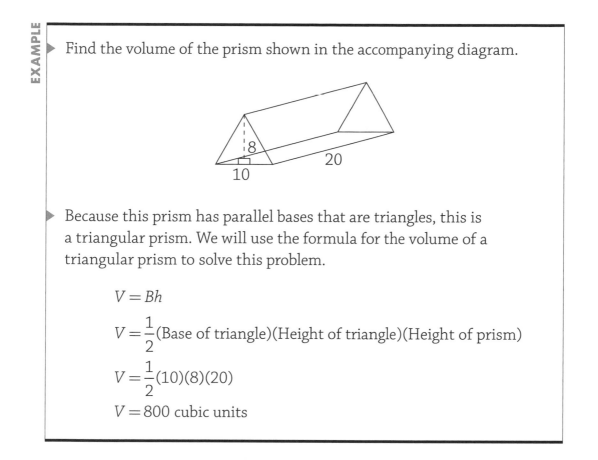

▶ Because this prism has parallel bases that are triangles, this is a triangular prism. We will use the formula for the volume of a triangular prism to solve this problem.

$$V = Bh$$

$$V = \frac{1}{2}(\text{Base of triangle})(\text{Height of triangle})(\text{Height of prism})$$

$$V = \frac{1}{2}(10)(8)(20)$$

$$V = 800 \text{ cubic units}$$

## EXTRA HELP

The following example will show how to find one of the dimensions of a prism when the volume is given.

**EXAMPLE**

▶ The volume of a rectangular prism is 495 in³. If the base of the prism is 9 in by 5 in, what is the height of the prism?

▶ Drawing a picture is helpful in solving these problems.

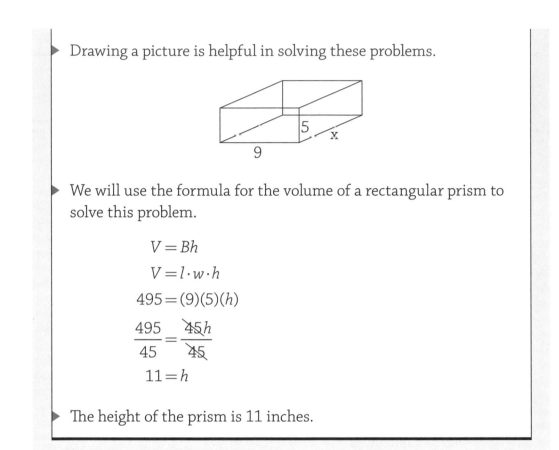

▶ We will use the formula for the volume of a rectangular prism to solve this problem.

$$V = Bh$$
$$V = l \cdot w \cdot h$$
$$495 = (9)(5)(h)$$
$$\frac{495}{45} = \frac{\cancel{45}h}{\cancel{45}}$$
$$11 = h$$

▶ The height of the prism is 11 inches.

## Square Pyramids

The next 3D figures we will discuss are **square pyramids**. A square pyramid is a solid object that has a base of a square and triangular sides that meet at the top, as shown in the accompanying diagram.

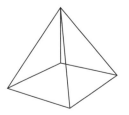

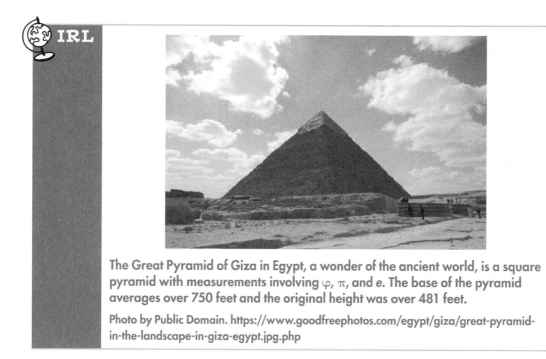

The Great Pyramid of Giza in Egypt, a wonder of the ancient world, is a square pyramid with measurements involving $\varphi$, $\pi$, and $e$. The base of the pyramid averages over 750 feet and the original height was over 481 feet.

Photo by Public Domain. https://www.goodfreephotos.com/egypt/giza/great-pyramid-in-the-landscape-in-giza-egypt.jpg.php

The formula for the volume of a pyramid is as follows:

$V = \dfrac{1}{3}Bh$, where $B$ is the area of the base of the pyramid and $h$ is the height of the pyramid.

Let's look at a problem that requires us to use this formula.

**EXAMPLE**

▶ A pyramid that is 10 inches tall has a square base whose length is 15 inches. What is the volume of the pyramid?

▶ Because the base of this pyramid is a square, $B$ will be the formula for the area of a square, which is $l \times w$.

$$V = \frac{1}{3}Bh$$

$$V = \frac{1}{3}l \cdot w \cdot h$$

$$V = \frac{1}{3}(15)(15)(10)$$

▶ The volume of the pyramid is 750 in$^3$.

## Spheres

A **sphere** is a perfectly round 3D shape. We see spherical objects often in everyday life. An inflated soccer ball, a globe in your social studies classroom, and a model of the planet Venus are all objects that can be classified as spheres. The volume of a sphere is found by using the following formula:

$$V = \frac{4}{3}\pi r^3,$$ where $r$ represents the radius of the sphere.

Let's use this formula to find the volume of a sphere given the length of the radius.

The accompanying diagram shows a sphere with a 5-inch radius. What is the volume of the sphere? Round your answer to the nearest tenth of a cubic inch.

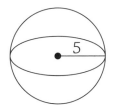

To solve this problem, all we need to do is plug our information into the formula for the volume of a sphere.

$$V = \frac{4}{3}\pi r^3$$

$$= \frac{4}{3}\pi(5)^3$$

$$= 523.6 \text{ in}^3$$

**EASY MISTAKE**

Never round until you have reached your final answer. When working with pi, it is always best to keep your answer in terms of pi until you have reached your last step of work and are ready to round.

The next example will show how to solve for the diameter of a sphere when the volume is given.

EXAMPLE

▶ What is the diameter of an inflated round ball that has a volume of $36\pi$ cubic inches?

▶ The only two variables in the formula for the volume of a sphere are the $V$, which represents the volume, and the $r$, which represents the radius. In this problem we were given the volume, which means we will be able to find the radius.

$$V = \frac{4}{3}\pi r^3$$

$$\left(\frac{3}{4}\right)36\pi = \frac{\cancel{4}}{\cancel{3}}\pi r^3\left(\frac{\cancel{3}}{\cancel{4}}\right)$$

$$\frac{27\cancel{\pi}}{\cancel{\pi}} = \frac{\cancel{\pi}r^3}{\cancel{\pi}}$$

$$\sqrt[3]{27} = \sqrt[3]{r^3}$$

$$3 = r$$

▶ We know that a diameter is twice the radius. This means the diameter of this sphere is 2(3), or 6 inches.

## From 2D to 3D

Did you know that we can rotate 2D figures around lines and create 3D figures? Try to imagine what 3D figure will be formed when rectangle *ABCD* is rotated along one of its sides.

We will begin by drawing a picture of a rectangle.

It does not matter which side we rotate around, so let's imagine we were to rotate the rectangle around side $\overline{CD}$. This means that side $\overline{CD}$ will stay in place and side $\overline{AB}$ will rotate around it, as shown in the following diagram.

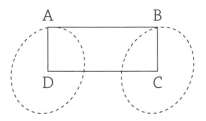

The resulting image will be a cylinder whose height is the length of side $\overline{AB}$ and circular bases whose radius is the length of side $\overline{BC}$, as shown in the following diagram.

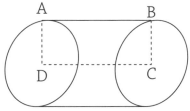

The following example will require you not only to determine the shape that will be created when rotating a figure but also to find the volume of the new shape.

EXAMPLE

▶ What is the volume of the figure created when rectangle *ABCD*, as shown in the accompanying diagram, is rotated around its shorter side? Leave your answer in terms of π.

▶ We know that opposite sides of a rectangle are congruent. Because we are rotating around the shortest side, we can rotate around either $\overline{AB}$ or $\overline{CD}$. The following diagram shows rectangle $ABCD$ rotated around side $\overline{CD}$.

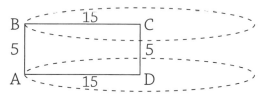

▶ This results in a cylinder whose height is the length of side $\overline{AB}$ and whose radius is the length of side $\overline{BC}$.

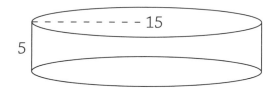

▶ We now need to plug our information into the formula for the volume of a cylinder.

$$V = Bh$$
$$= \pi r^2 h$$
$$= \pi (15)^2 (5)$$
$$= 1{,}125\pi \text{ cubic units}$$

The following example applies this concept toward a different 2D figure.

▶ The accompanying diagram shows right triangle $ABC$, where $AB = 12$ and $BC = 5$. Determine the 3D figure created when triangle $ABC$ is rotated around side $\overline{AB}$. Find the volume of the 3D figure to the nearest hundredth.

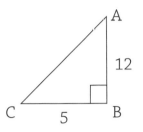

▶ Because we are rotating around side $\overline{AB}$, this means $\overline{AB}$ will not move but $\overline{BC}$ and $\overline{AC}$ will rotate around it as shown in the accompanying diagram.

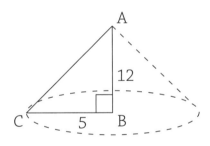

▶ The resulting 3D figure will therefore be a cone whose base is a circle with a radius of length $\overline{BC}$ and whose height is the length of side $\overline{AB}$ as shown in the following diagram.

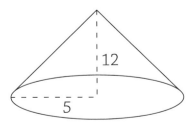

▶ We now need to plug our information into the formula for the volume of a cone.

$$V = \frac{1}{3}Bh$$

$$= \frac{1}{3}\pi r^2 h$$

$$= \frac{1}{3}\pi(5)^2(12)$$

$$= 314.16 \, \text{unit}^3$$

Let's try this concept with one more example.

▶ The following diagram shows $\overset{\frown}{AB}$ with point $A(-3, 0)$ and point $B(0, 3)$. Determine the 3D figure formed when $\overset{\frown}{AB}$ is rotated around the $y$-axis. Find the volume of the figure in terms of pi.

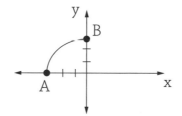

▶ Because the figure is being rotated around the $y$-axis, point $B$ will not move but point $A$ will rotate around the $y$-axis, as shown in the accompanying diagram.

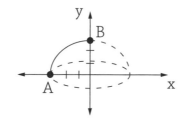

▶ The resulting 3D figure will be a hemisphere (half a sphere) with a radius the length of $\overline{OA}$, three units, as shown in the following diagram.

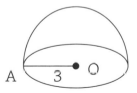

▶ Now we need to plug our information into the formula for the volume of a sphere.

$$V = \frac{4}{3}\pi r^3$$

$$= \frac{4}{3}\pi(3)^3$$

$$= 36\pi \text{ units}^3$$

▶ This is the volume for the whole sphere. Because we only want the volume of half of a sphere, we need to divide our answer by 2. This means that the volume of our 3D figure is $18\pi$ units$^3$.

## BTW

*Cavalieri's Principle* states that if two different 3D figures have the same height, and all cross sections drawn parallel to the bases from the same height have the same area, then the volumes of the figures are equal. The following diagram shows this principle using a stack of quarters.

*Both stacks have the same base, the same height, and the same cross section throughout the entire height. This means that the volumes of both stacks of quarters are equal.*

## EXERCISES

### EXERCISE 13-1

*Solve the following problems.*

1. The radius of the base of a cylinder is 20 inches and the height is 24 inches. Find the surface area of the cylinder in terms of $\pi$.

2. The diameter of the base of a cylindrical above-ground pool is 20 feet. The height of the pool is 7 feet. Find the volume of the pool to the nearest tenth of a cubic foot.

3. The volume of a cylinder is $100\pi$. The radius is 5. Find the height of the cylinder.

4. The local ice cream shop sells ice cream cones that have a height of 5 inches and a radius of 2.5 inches. How much ice cream can be scooped and put inside a cone? Round the answer to the nearest tenth of a cubic inch.

5. A soccer coach is using small traffic cones to run his dribbling drills. Each cone has a radius of 3 inches and a height of 4 inches. Find the slant height of the cone.

6. The slant height of a cone is 25 centimeters. If the radius is 7 centimeters, what is the height of the cone?

7. Find the volume of a square pyramid where the length of its base is 4 meters and its height is 9 meters.

8. Find the volume of a 6-inch-high prism whose base is a rectangle measuring 10 inches by 3 inches.

9. Find the volume of the following prism.

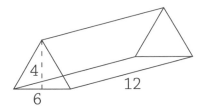

10. The diameter of a fully inflated round ball is 7 centimeters. Find the volume of the ball to the nearest hundredth of a cubic centimeter.

11. Find the radius of a fully inflated round ball that has a volume of 12 cm³ to the nearest hundredth of a centimeter.

## EXERCISE 13-2

***Using the given information, determine the volume described in these problems.***

1. Find the volume of the 3D figure formed when the rectangle in the accompanying diagram is rotated around side $\overline{BC}$. Leave your answer in terms of $\pi$.

2. Rectangle *ABCD* has sides with the following measurements: $AB = 12$ cm and $BC = 20$ cm. Find the volume of the 3D figure created when rectangle *ABCD* is rotated around the perpendicular bisector of $\overline{BC}$. Round your answer to the nearest tenth of a cubic centimeter.

3. Find the volume of the 3D figure created when the accompanying diagram of triangle $ABC$ is rotated around side $\overline{BC}$. Round your answer to the nearest tenth of a cubic unit.

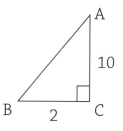

4. Find the volume of the 3D figure created when the accompanying diagram of semicircle $O$ is rotated around diameter $\overline{AOB}$, which is 12 inches long. Round your answer to the nearest hundredth of a cubic inch.

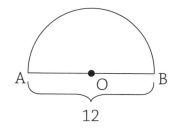

Flashcard App

# 14 Constructions

## MUST KNOW

⚡ With just a straight edge and a compass, we can construct many geometric shapes.

⚡ Constructions provide a visual representation of many geometric properties, including angle and perpendicular bisectors, centers of triangles, and transformation properties.

any of the geometric properties we have learned can be shown through constructions. In this chapter, you will be given step-by-step instructions on the most commonly performed geometric constructions. To be able to perform these constructions, you will need to have your compass handy.

## Copying Segments and Angles

Let's begin by learning how to **copy a line segment**. When we copy a line segment, we are constructing a line segment congruent to a given line segment. The following diagram shows line segment $\overline{AB}$. We are going to construct $\overline{CD}$, a line segment congruent to $\overline{AB}$.

A                    B

**Step 1**  We begin by taking a straight edge and drawing a line segment significantly longer than $\overline{AB}$ and labeling the left endpoint $C$.

A              B

C

**Step 2**  Place your compass point at point $A$ and open the compass until the pencil touches point $B$.

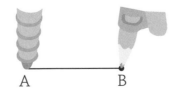

A              B

> **EASY MISTAKE**
>
> Using a compass can be quite challenging to some people. Make sure you hold the compass in place with one hand and you gently spin the compass to make your arcs.

**Step 3** Without changing the width of the compass, place the compass point at point $C$ and make an arc through the line you drew in step 1.

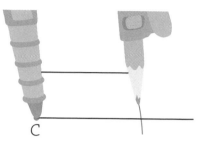

C

**Step 4** Label the intersection point $D$. You have now copied $\overline{AB}$. We can now state that $\overline{CD} \cong \overline{AB}$.

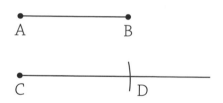

In the next construction, we will learn to **copy a given angle**. The following diagram shows $\angle ABC$. We will be constructing $\angle DEF$, an angle congruent to $\angle ABC$.

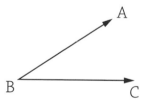

**Step 1** We begin by taking a straight edge and drawing a line segment and labeling the left endpoint $E$, the vertex of the new angle.

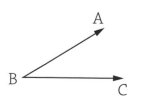

**Step 2** Placing the compass point at $B$, open the compass wide enough so that you can make an arc that intersects $\angle ABC$ on both rays.

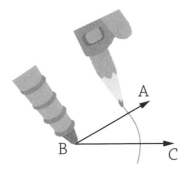

**Step 3** Without changing the width of the compass, place the compass point at $E$. Draw a generous-sized arc through the line segment and label the intersection point $F$.

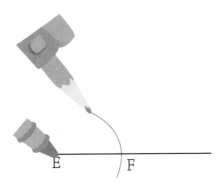

**Step 4** Go back to $\angle ABC$ and place the compass point and the pencil at the locations where the arc intersected the rays.

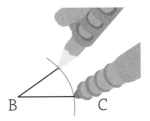

**Step 5** Without changing the width of the compass, place the compass point at $F$ and draw an arc that intersects the arc you drew in step 3. Label the intersection point $D$.

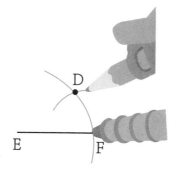

**Step 6** Using a straight edge, draw a ray from point $E$ through point $D$. You have now copied $\angle ABC$. We can now state that $\angle DEF \cong \angle ABC$.

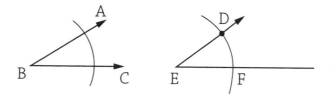

## Bisectors and Perpendicular and Parallel Lines

We will now learn how to **construct an angle bisector**. An angle bisector divides an angle into two congruent angles. The following diagram shows $\angle ABC$. We will be constructing $\overrightarrow{BG}$, the bisector of $\angle ABC$.

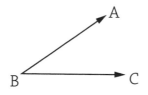

**Step 1** Placing the compass point at $B$, open the compass wide enough so that you can make an arc that intersects $\angle ABC$ on both rays.

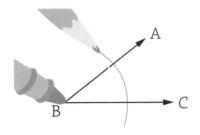

**Step 2** Place the compass point at the location where the arc intersected $\overrightarrow{BC}$ and draw a generous arc.

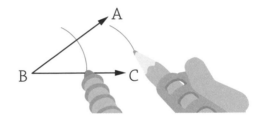

**Step 3** Place the compass point at the location where the arc intersected $\overrightarrow{BA}$ and draw a generous arc. This arc should intersect the arc drawn in step 2. Label the intersection of these arcs point $G$.

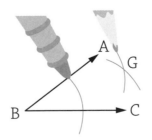

**Step 4** Using a straight edge, draw a ray from point $B$ through point $G$. You have now constructed $\overrightarrow{BG}$, the bisector of $\angle ABC$. We can now state that $\angle ABG \cong \angle CBG$.

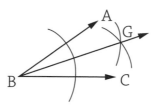

# Constructions Involving Perpendicular Lines

The next several constructions will focus on perpendicular lines. First, we will **construct a line perpendicular to a given line through a given point**. The following diagram shows $\overline{AEB}$. We will be constructing $\overline{DEF}$, a line segment perpendicular to $\overline{AEB}$ at point $E$.

**Step 1**    Place the compass point at $E$ and draw an arc that intersects $\overline{AB}$ on both sides of point $E$.

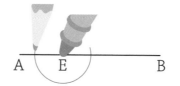

**Step 2**    Place the compass point at the first location where the arc intersected $\overline{AB}$ and draw a generous arc. It does not matter if you draw the arc above $\overline{AB}$ or below $\overline{AB}$.

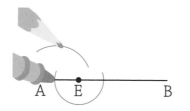

**Step 3** Without changing the width of the compass, place the compass point at the second location where the arc intersected $\overline{AB}$ and draw a generous arc. This arc should intersect the arc drawn in step 2. Label the intersection point $D$.

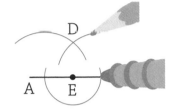

**Step 4** Using a straight edge, draw a line segment from point $D$ through point $E$ and label the endpoint $F$. You have now constructed $\overline{DEF}$, a line segment perpendicular to $\overline{AEB}$ at point $E$. We can now state that $\overline{DF} \perp \overline{AB}$ and $\angle DEA \cong \angle DEB$. This also means that $\angle DEA$ and $\angle DEB$ are right angles.

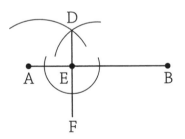

## EXTRA HELP

The construction of a perpendicular line through a point is the same whether the given point is on or off the line. The following diagram shows the construction of $\overline{EF}$ perpendicular to $\overline{AB}$ with point $E$ not on $\overline{AB}$. Notice we decided to make our arcs below $\overline{AB}$ instead of above. We did this to show you that it does not matter where you choose to draw the arcs.

(continued on next page)

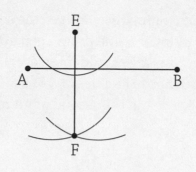

The final construction that involves perpendicular lines will be the **construction of a perpendicular bisector**. If you remember from our geometric properties, this will be a line or line segment perpendicular to a given line segment that intersects the given line segment at its midpoint. We perform this construction if we are asked to bisect a line segment whether or not the word *perpendicular* is mentioned. The following diagram shows $\overline{AB}$. We will be constructing $\overleftrightarrow{CD}$, the perpendicular bisector of $\overline{AB}$.

**Step 1**   Place the compass point at point $A$ and open the compass so that its width is more than half the width of $\overline{AB}$. Draw a generous arc above and below $\overline{AB}$.

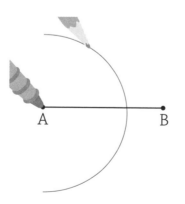

**Step 2** Without changing the width of the compass, place the compass point at $B$ and draw a generous arc above and below $\overline{AB}$. This arc should intersect the arc drawn in step 1 both above and below $\overline{AB}$. Label the first point of intersection $C$ and the second point of intersection $D$.

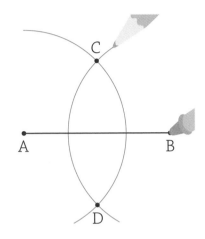

**Step 3** Using a straight edge, connect points $C$ and $D$. You have now constructed $\overline{CD}$, the perpendicular bisector of $\overline{AB}$. Notice that the point where $\overline{CD}$ intersects $\overline{AB}$ is the midpoint of $\overline{AB}$.

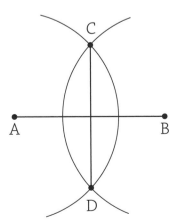

## Constructing Parallel Lines

We have practiced all the perpendicular constructions, which means it is time to learn how to **construct a line parallel to a given line that passes through a given point**. The following diagram shows $\overline{AB}$ with point $C$ not on $\overline{AB}$. We will be constructing a line parallel to $\overline{AB}$ passing through point $C$.

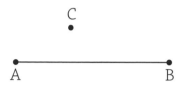

**Step 1**   Using a straight edge, draw a line segment through point $C$ that intersects $\overline{AB}$ at an angle.

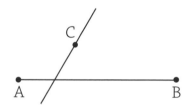

**Step 2**   Place your compass at the point where your line segment intersected $\overline{AB}$. Open the compass a small amount and draw an arc that intersects both line segments.

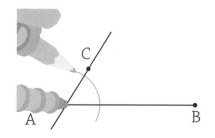

**Step 3**   Without changing the width of the compass, place the compass point at point $C$ and draw a generous arc that intersects the line segment drawn in step 1.

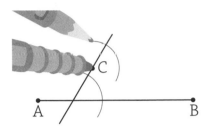

**Step 4**   Go back to the arc drawn in step 2 (the first arc drawn). Place the compass point and the pencil point at the two locations where the arc intersected the lines.

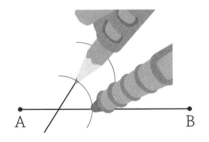

**Step 5**   Without changing the width of the compass, place the compass point at the location where the second arc drawn (the one from step 3) intersected the line segment drawn in step 1. Draw an arc. This arc should intersect the second arc. Label the point of intersection point $D$.

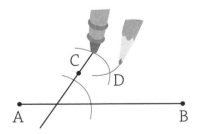

**Step 6**  Using a straight edge, draw a line connecting points $C$ and $D$. You have now constructed $\overleftrightarrow{CD}$ parallel to $\overline{AB}$.

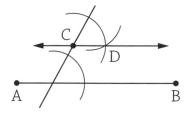

## Construction Applications

Now that we have gone over some basic geometric constructions, we can apply what we've learned toward some more intricate constructions. Let's begin by **constructing a triangle congruent to a given triangle**. Doing this requires the ability to copy a segment. The following diagram shows triangle $ABC$. We will be constructing $\triangle DEF$, a triangle congruent to $\triangle ABC$.

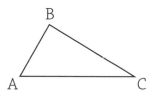

**Step 1**  Construct a line segment congruent to $\overline{AC}$. Label this segment $\overline{DF}$ because those are the endpoints that correspond with points $A$ and $C$ in the given triangle. If you forgot how to do this, see the beginning of the chapter and follow the steps for copying a line segment.

**Step 2** Place the compass point at point *A* and open it until the pencil meets point *B*. Without changing the width of the compass, place the compass point at *D* and draw a generous arc.

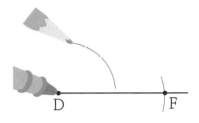

**Step 3** Place the compass point at point *C* and open it until the pencil meets point *B*. Without changing the width of the compass, place the compass at point *F* and draw a generous arc. This arc should intersect the arc drawn in step 2. Label the intersection point *E*, which represents the third vertex of the triangle.

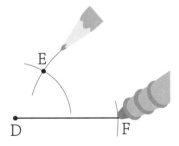

**Step 4** Using a straight edge, connect point *D* to point *E* and connect point *F* to point *E*. You have now constructed △*DEF* congruent to △*ABC*.

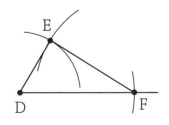

Let's apply these concepts toward **constructing an equilateral triangle**. An equilateral triangle has all sides congruent. The following diagram shows $\overline{AB}$. We will be constructing equilateral triangle $ABC$ with side length $\overline{AB}$.

**Step 1**   Place your compass point at point $A$ and open it until the pencil reaches point $B$. Draw a generous arc.

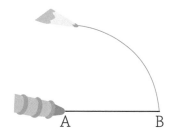

**Step 2**   Without changing the width of the compass, place the compass point at point $B$ and draw a generous arc. This arc should intersect the arc drawn in step 1. Label this intersection point $C$.

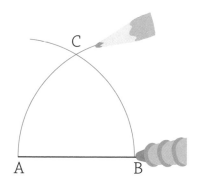

**Step 3** Using a straight edge, connect point $A$ to point $C$ and connect point $B$ to point $C$. You have now constructed an equilateral triangle with side lengths congruent to $\overline{AB}$.

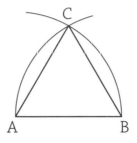

We already know that an equilateral triangle has three congruent sides. The angles of an equilateral triangle are also congruent, making each angle measure 60°. Constructing an equilateral triangle will be the first step for any construction that requires a 60° angle. For example, if you want to construct a 30° angle, you can construct a 60° angle and then bisect it into two equal 30° angles. The following steps show the **construction of a 30° angle**.

**Step 1** Using a straight edge, draw a segment of any length.

_____

**Step 2** Construct a 60° angle. If you forgot how to do this, follow the steps for constructing an equilateral triangle.

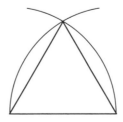

**Step 3** Construct an angle bisector through any of the three 60° angles constructed in step 2. We will be constructing our angle bisector through the bottom left angle of the equilateral triangle. If you forgot how to do this, look back in the chapter for the steps on constructing an angle bisector. You have now constructed two 30° angles.

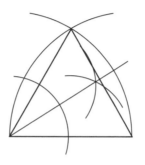

Can you think of any previous constructions we have learned that will help us construct a 45° angle? If bisecting a 60° angle helped us construct a 30° angle, then bisecting a 90°angle will help us **construct a 45° angle**. How can we construct a 90° angle? You guessed it! Construct perpendicular lines! Try it below.

**Step 1** Using a straight edge, draw a line segment of any length.

--------------------

**Step 2** Construct a line perpendicular to the line segment drawn in step 1. It is probably easiest to construct a perpendicular bisector but you could technically draw a line perpendicular to any point along the given segment.

If you forgot how to construct a perpendicular bisector, look back in the chapter and follow the steps. You have now created four 90° angles.

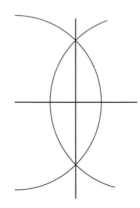

**Step 3** Construct an angle bisector through any of the four 90° angles you constructed in step 2. We will construct our angle bisector through the top left 90° angle. If you forgot how to do this, go back in the chapter and follow the steps for constructing an angle bisector. You have now created two 45° angles.

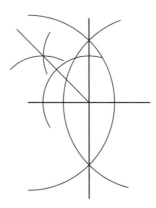

## Constructing an Altitude and a Median

An *altitude* is a line segment drawn from the vertex of a triangle perpendicular to the line containing the opposite side. Because we have already learned how to construct perpendicular lines, we know enough to **construct the altitude of a triangle**. The following diagram shows $\triangle ABC$. We will be constructing $\overline{AD}$, the altitude of triangle *ABC*, from vertex *A* to side $\overline{BC}$.

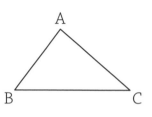

> **EASY MISTAKE**
>
> Not all altitudes intersect the actual side opposite the vertex. There are instances where in order to construct an altitude you will have to extend a side of the triangle.

**Step 1**   Construct a line perpendicular to $\overline{BC}$ going through point *A*. If you forgot how to do this, go back in the chapter and follow the steps for constructing a perpendicular line through a given point. Label the point where the perpendicular line intersects side $\overline{BC}$ point *D*. You have now constructed $\overline{AD}$, the altitude from vertex *A* to side $\overline{BC}$.

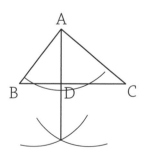

A median is a line segment drawn from the vertex of a triangle to the midpoint of the opposite side. We previously learned that a perpendicular bisector will divide a segment into two equal segments at the midpoint.

This means we know enough information to construct a median.
The following diagram shows △*ABC*. We will be constructing $\overline{AD}$, **the median of triangle *ABC***, from vertex *A* to side $\overline{BC}$.

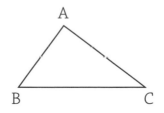

**Step 1**  Construct the perpendicular bisector of $\overline{BC}$ and label the midpoint *D*. If you forgot how to do this, go back in the chapter and follow the steps for constructing a perpendicular bisector.

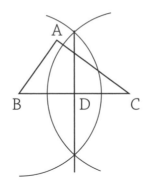

**Step 2**  Using a straight edge, connect point *A* to point *D*. You have now constructed $\overline{AD}$, the median of triangle *ABC* from vertex *A* to side $\overline{BC}$.

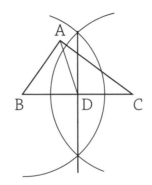

## Constructing a Square and Hexagon Inscribed in a Circle

The next two constructions to be discussed will involve constructing shapes other than triangles. Our first construction will be **a square inscribed in a circle**, and the second construction will be a regular hexagon. Let's start with constructing the square inscribed in circle $O$.

**Step 1**    Place point $O$ anywhere on your paper. With your compass open to any width, draw a circle.

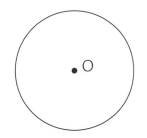

**Step 2**    Using your straight edge, draw a diameter. Remember that the diameter of a circle is a chord that goes through the center of the circle.

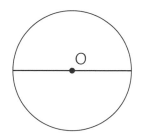

**Step 3** Construct a perpendicular bisector through the diameter you drew in step 2. Be sure to extend the perpendicular bisector to intersect the circle twice.

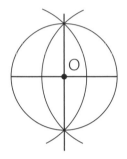

**Step 4** Using a straight edge, connect the four points where the endpoints intersected the circle to form the square.

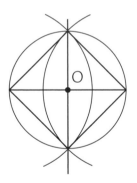

Now let's learn how to construct a regular hexagon inscribed in circle O.

**Step 1** Place point O anywhere on your paper. With your compass open to any width, draw a circle.

**Step 2** Place a point anywhere along circle O. Label that point A. Without changing the width of your compass from step 1, place your compass point on point A and draw an arc intersecting the circle. Label that point B.

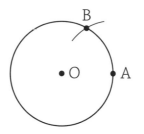

**Step 3** Keeping the width of the compass the same, place the compass point at point B and draw another arc through the circle. Label that point C.

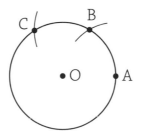

Continue this process until you have five arcs drawn on the circle.

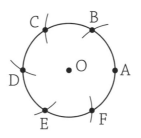

**Step 4**   Using a straight edge, connect the six points along the circle to form hexagon *ABCDEF*.

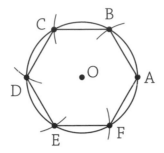

## Constructing Transformations

This section will discuss constructions involving transformations. We will show the constructions of line reflections, rotations, translations, and dilations. We will first learn how to **construct the rotation of a point**. The following diagram shows $\overline{AB}$ and $\angle DEF$. We will be constructing $B'$, the rotation of point $B$ around point $A$ using $\angle DEF$.

**Step 1**   Place the compass point at point *A* and the pencil at point *B* and draw a circle.

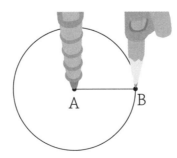

**Step 2**   Copy ∠*DEF* using $\overline{AB}$ as part of the angle. If you forgot how to do this, go back to the beginning of the chapter and follow the steps for copying an angle.

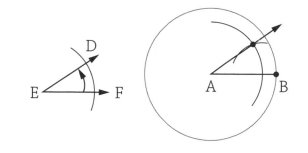

**Step 3**   Label the location where the new ray intersects the circle *B'*. You have now rotated point *B* around point *A* using ∠*DEF*.

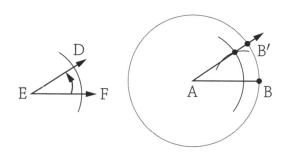

In the preceding example, we learned how to rotate about a point. This point is called the *center of rotation*. We will now learn how to **construct the center of rotation**. The accompanying diagram shows △*A'B'C'*, the image of

$\triangle ABC$ after a rotation around point $P$. We will be constructing point $P$, the center of rotation.

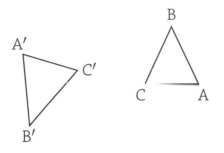

**Step 1**  Connect a point to its image point and construct a perpendicular bisector. We will connect $C$ to $C'$. Look at the beginning of the chapter and follow the steps for the construction of a perpendicular bisector if you forgot how to do this.

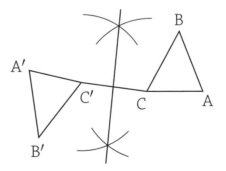

**Step 2** Repeat step 1 with another pair of image points. We will use $B$ and $B'$.

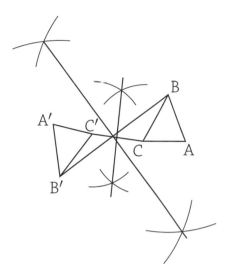

**Step 3** Label the point where the two perpendicular bisectors drawn in steps 1 and 2 intersected point $P$. You have now constructed the center of rotation.

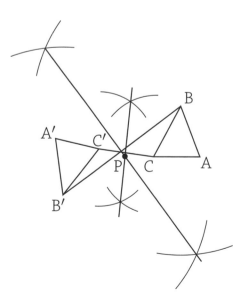

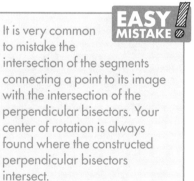

EASY MISTAKE

It is very common to mistake the intersection of the segments connecting a point to its image with the intersection of the perpendicular bisectors. Your center of rotation is always found where the constructed perpendicular bisectors intersect.

The next transformation we will discuss will be a reflection. The following diagram shows $\overline{AB}$ and line $l$. We will be **constructing $\overline{A'B'}$, the image of $\overline{AB}$ after a reflection over line $l$.**

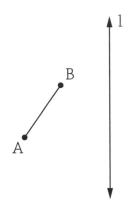

**Step 1** Construct a line perpendicular to line $l$ that passes through point $B$. Be sure not to change the width of the compass when you are doing your construction. Label the point where the arcs intersect $B'$. Go back to the beginning of the chapter and follow the steps for constructing a perpendicular line through a given point if you forgot how to do this step.

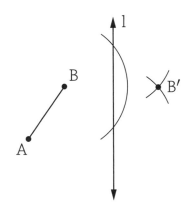

**Step 2**  Repeat step 1 using point *A* instead of point *B*. Label the point where the arcs intersect, point *A'*.

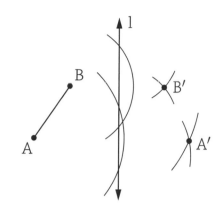

**Step 3**  Using a straight edge, connect *A'* and *B'*. You have now constructed $\overline{A'B'}$, the image of $\overline{AB}$ after a reflection over line *l*.

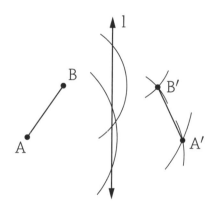

If we can use a compass to construct the reflection of points over a line, we can also use a compass to **construct a line of reflection**. The following diagram below shows $\overline{A'B'}$ the image of $\overline{AB}$ after a reflection over line *l*. We will be constructing line *l*, the line of reflection.

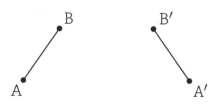

**Step 1** Using a straight edge, connect any point to its image point to create a segment. In this diagram, we connected $A$ to $A'$.

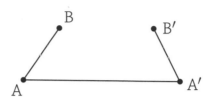

**Step 2** Construct the perpendicular bisector of the segment drawn in step 1 and label it $l$. You have now constructed line $l$, the line of reflection. If you forgot how to do this step, go back to the beginning of the chapter and follow the steps for constructing a perpendicular bisector.

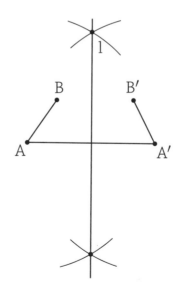

Dilations are another transformation that we can construct using a compass. The diagram below shows △*ABC*. We will be constructing △*A′B′C′*, the image of △*ABC* after a dilation with a scale factor of 2 and the center of dilation at point *A*.

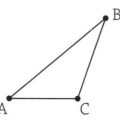

**Step 1** Using a straight edge, extend side $\overline{AC}$ from point *A* through point *C*. Extend side $\overline{AB}$ from point *A* through point *B*.

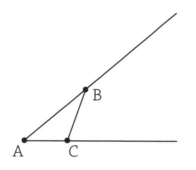

**Step 2** Use the compass to measure the length of $\overline{AC}$ by placing the compass point at point *A* and the pencil at point *C*. Without changing the width of the compass, place the compass point at point *C* and make an arc through the extended line segment. Label this point *C′*.

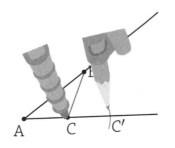

**Step 3** Repeat step 2 for $\overline{AB}$ by placing the compass point at $A$ and the pencil at point $B$. Then, without changing the width of the compass, place the compass point at point $B$ and make an arc through the extended line segment. Label that point $B'$.

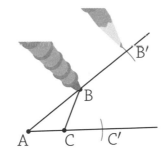

**Step 4** Using a straight edge, connect $B'$ to $C'$. You have now constructed $\triangle A'B'C'$, the image of $\triangle ABC$ after a dilation with a scale factor of 2 and a center of dilation at point $A$.

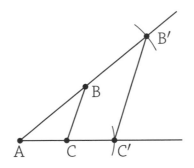

The last transformation we will be constructing in this chapter will be translations. When we **construct a translation**, we will be given a vector for the translation. The vector shows the magnitude and direction of the translation. This means that it shows the length and direction of the translation. The following diagram shows $\overline{AB}$ and vector $\overrightarrow{CD}$. We will be constructing $\overline{A'B'}$, the image of $\overline{AB}$ after a translation along vector $\overrightarrow{CD}$.

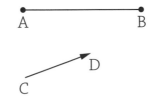

**Step 1** Place the compass point at $C$ and the pencil at $D$ to measure the amount point $A$ needs to shift. Without changing the width of the compass, place the compass point at $A$ and make a generous arc.

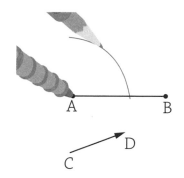

**Step 2** The distance from $C$ to $A$ must be the same as the distance from $D$ to $A'$. Place the compass point at $C$ and the pencil at $A$ to measure the distance from point $C$ to point $A$. Without changing the width of the compass, place the compass point at $D$ and make an arc that intersects the arc that was previously drawn. Label the intersection point $A'$.

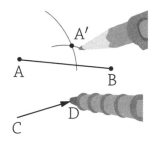

We will now repeat steps 1 and 2 for point $B$.

**Step 3** Place the compass point at $C$ and the pencil at $D$ to measure the distance it needs to shift. Without changing the width of the compass, place the compass point at $B$ and make a generous arc.

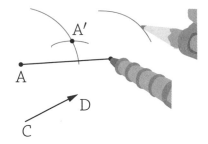

**Step 4** Place the compass point at $C$ and the pencil at $B$ to measure the distance from point $C$ to point $B$. Without changing the width of the compass, place the compass point at $D$ and make an arc that intersects the arc that was previously drawn. Label the intersection point $B'$.

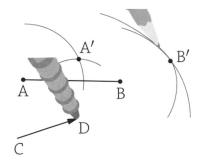

**Step 5** Using a straight edge, connect $A'$ to $B'$. You have now constructed $\overline{A'B'}$, the image of $\overline{AB}$ after a translation along vector $\overrightarrow{CD}$.

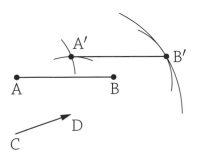

## EXERCISES

### EXERCISE 14-1

*Follow the instructions for each problem.*

1.  Construct a perpendicular bisector to $\overline{AB}$.

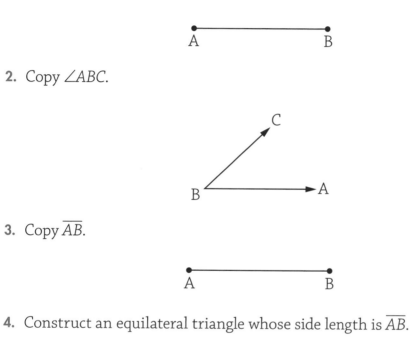

A ————————— B

2.  Copy $\angle ABC$.

3.  Copy $\overline{AB}$.

A ————————— B

4.  Construct an equilateral triangle whose side length is $\overline{AB}$.

A——————————B

5.  Construct the bisector of $\angle ABC$.

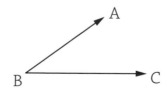

6. Construct a line perpendicular to $\overline{AB}$ through point $C$.

7. Construct a hexagon inscribed in a circle.

8. Construct $\overline{BD}$, the median of $\triangle ABC$.

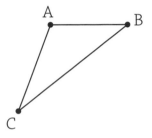

9. Rotate point $P$ 60° clockwise around point $C$. *Hint:* You will first need to construct a 60° angle.

10. What does the following diagram show a construction of?

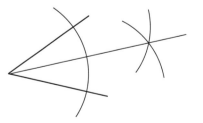

11. What does the following diagram show a construction of?

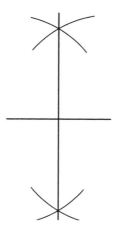

12. Construct the line of reflection for △ABC and its image △A′B′C′.

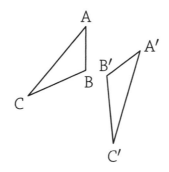

13. Describe how you could construct a 45° angle.

**14.** Copy △ABC.

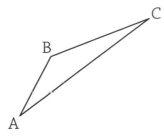

**15.** Dilate $\overline{AB}$ by a scale factor of 2 with the center of dilation at point C.

# Answer Key

## 1

## Definitions

1. **30°**   A bisector divides an angle into two congruent angles. $\overline{BD}$ bisects $\angle ABC$; therefore

$$m\angle ABD = \frac{1}{2}m\angle ABC$$

$$m\angle ABD = \frac{1}{2}(60°)$$

$$m\angle ABD = 30°$$

2. **20**   A bisector divides a segment into two congruent segments. $\overline{AB}$ bisects $\overline{CD}$ at $E$; therefore

$$CD = 2(CE)$$
$$CD = 2(10)$$
$$CD = 20$$

3. **40**   A bisector divides a segment into two congruent segments. $\overline{AB}$ bisects $\overline{CD}$ at $E$; therefore

$$CE = ED$$
$$\cancel{3x} + 5 = 7x - \cancel{15}$$
$$\underline{-\cancel{3x} + 15 \quad -3x + \cancel{15}}$$
$$\frac{20}{4} = \frac{\cancel{4x}}{\cancel{4}}$$
$$5 = x$$
$$CE + ED = CD$$
$$3x + 5 + 7x - 15 = CD$$
$$3(5) + 5 + 7(5) - 15 = CD$$
$$40 = CD$$

4. **15** Intersecting lines form congruent vertical angles. $\overline{AB}$ and $\overline{CD}$ intersect at E; therefore

$$m\angle CEB = m\angle AED$$
$$2x = \cancel{x} + 15$$
$$\underline{-x \quad -\cancel{x}}$$
$$x = 15$$

5. **40** Perpendicular lines form right angles. Right angles equal 90 degrees. $\overline{AB} \perp \overline{CD}$ at E; therefore

$$m\angle AED = 90°$$
$$2x + \cancel{10} = 90°$$
$$\underline{-\cancel{10} \quad -10}$$
$$\frac{\cancel{2}x}{\cancel{2}} = \frac{80}{2}$$
$$x = 40$$

## EXERCISE 1-2

1. **65°**    The sum of two complementary angles is 90°. $\angle A$ and $\angle B$ are complementary angles; therefore

$$m\angle A + m\angle B = 90°$$
$$\cancel{25}° + m\angle B = 90°$$
$$\underline{-\cancel{25}° \qquad\quad -25°}$$
$$m\angle B = 65°$$

2. **85°**    The sum of two supplementary angles is 180°. $\angle C$ and $\angle D$ are supplementary angles; therefore

$$m\angle C + m\angle D = 180°$$
$$\cancel{95}° + m\angle D = 180°$$
$$\underline{-\cancel{95}° \qquad\quad -95°}$$
$$m\angle D = 85°$$

## EXERCISE 1-3

1. **10**    Adjacent angles that form a linear pair add to 180°. $\angle ABC$ and $\angle CBD$ form a linear pair; therefore

$$m\angle ABC + m\angle CBD = 180°$$
$$12x + 5 + 8x - 25 = 180$$
$$20x - \cancel{20} = 180$$
$$\underline{+\cancel{20} \quad +20}$$
$$\frac{\cancel{20}x}{\cancel{20}} = \frac{200}{20}$$
$$x = 10$$

2. **10**   A midpoint divides a segment into two congruent segments. $A$ is the midpoint of $\overline{CT}$; therefore

$$AT = CA$$
$$AT = 2x - 4$$
$$CA + AT = CT$$
$$2x - 4 + 2x - 4 = 3x + 2$$
$$4x - 8 = 3x + 2$$
$$\underline{-3x + 8 \quad -3x + 8}$$
$$x = 10$$

3. **20°**   A bisector divides an angle into two congruent angles. $\overline{GE}$ bisects $\angle DEG$; therefore

$$m\angle DEF = m\angle FEG$$
$$3x + 2 = 5x - 10$$
$$\underline{-3x + 10 \quad -3x + 10}$$
$$\frac{12}{2} = \frac{2x}{2}$$
$$6 = x$$
$$m\angle DEG = 3x + 2$$
$$m\angle DEG = 3(6) + 2$$
$$m\angle DEG = 20°$$

4. $\overline{MA} \cong \overline{AP}$, $\angle SAM \cong \angle SAP \cong \angle TAM \cong \angle TAP$   Perpendicular lines form congruent right angles. A bisector divides a segment into two congruent segments.

5. **Reflexive property**   The reflexive property states that a segment is always congruent to itself.

6. $\angle PQR$   The sum of two angles that share a common ray is a larger angle with a vertex at the same letter.

7. $\overline{RO}$   If you subtract the same segment from two congruent segments, the resulting segments are congruent.

## EXERCISE 1-4

| Statement | Reason |
|---|---|
| 1) $\overline{ABLE}$, $\overline{AB} \cong \overline{EL}$ | 1) Given |
| 2) $\overline{BL} \cong \overline{BL}$ | 2) Reflexive property |
| 3) $\overline{AL} \cong \overline{EB}$ | 3) Addition postulate |

# 2
# Triangle Proofs

## EXERCISE 2-1

1. **SAS $\cong$ SAS**   If two sides and the included angle of one triangle are congruent to two sides and the included angle of another triangle, then the triangles are congruent.

2. **AAS $\cong$ AAS**   If two angles and the non-included side of one triangle are congruent to two angles and the non-included side of another triangle, then the triangles are congruent.

3. **HL $\cong$ HL**   If the hypotenuse and leg of a right triangle are congruent to the hypotenuse and leg of another right triangle, then the right triangles are congruent.

4. **SSS $\cong$ SSS**   If three sides of one triangle are congruent to three sides of another triangle, then the triangles are congruent.

## EXERCISE 2-2

1. $\overline{RT} \cong \overline{XT}$   If two angles and the included side of one triangle are congruent to two angles and the included side of another triangle, then the triangles are congruent. $\overline{RT}$ and $\overline{XT}$ are the included sides between the congruent angles in the triangles.

2. $\overline{TS} \cong \overline{TW}$ or $\overline{RS} \cong \overline{XW}$   If two angles and the non-included side of one triangle are congruent to two angles and the non-included side of another triangle, then the triangles are congruent. $\overline{TS}$ and $\overline{TW}$ are a pair of corresponding non-included sides of the triangle. $\overline{RS}$ and $\overline{XW}$ are another pair of corresponding non-included sides of the triangle.

## EXERCISE 2-3

The accompanying table shows the statements and reasons that lead to the triangles being congruent because of the angle-side-angle postulate.

| Statement | Reason |
|---|---|
| $\angle A \cong \angle E$ | Given |
| $\overline{AD} \cong \overline{EC}, \overline{BD} \cong \overline{BC}$ | 1. **Given** |
| 2. $\overline{AB} \cong \overline{EB}$ | Addition postulate |
| 3. $\angle B \cong \angle B$ | Reflexive property |
| $\triangle ABC \cong \triangle EBD$ | 4. **ASA $\cong$ ASA** |

## EXERCISE 2-4

The accompanying table shows the statements and reasons that lead to the triangles being congruent because of the angle-angle-side postulate. If two triangles are congruent, then their corresponding sides are congruent.

| Statement | Reason |
|---|---|
| $\overline{MR} \cong \overline{TP}$ | Given |
| $\overline{RP} \cong \overline{RP}$ | 1. **Reflexive property** |
| $\overline{MP} \cong \overline{TR}$ | 2. **Addition postulate** |
| $\triangle RAP$ is isosceles | Given |

| Statement | Reason |
|---|---|
| 3. $\angle ARP \cong \angle APR$ | If a triangle is isosceles, then its base angles are congruent. |
| $\overline{MN} \perp \overline{PN}$ <br> $\overline{TS} \perp \overline{RS}$ | Given |
| $\angle N$ and $\angle S$ are right angles. | Perpendicular lines form right angles. |
| $\angle N \cong \angle S$ | Right angles are congruent. |
| 4. $\triangle MNP \cong \triangle TSR$ | AAS $\cong$ AAS |
| $\overline{MN} \cong \overline{TS}$ | 5. **Corresponding parts of congruent triangles are congruent (CPCTC)** |

## EXERCISE 2-5

1. **$x = 10$**    If the triangles are congruent, then their corresponding angles are congruent. $\angle C$ and $\angle T$ are corresponding angles; therefore

$$m\angle C = m\angle T$$
$$\cancel{2}x + 40 = 8x - \cancel{20}$$
$$\underline{-\cancel{2}x + 20 \quad -2x + \cancel{20}}$$
$$\frac{60}{6} = \frac{\cancel{6}x}{\cancel{6}}$$
$$x = 10$$

2. **$x = 2$**    If the triangles are congruent, then their corresponding sides are congruent. $\overline{AC}$ and $\overline{RT}$ are corresponding sides; therefore

$$AC = RT$$
$$14x - \cancel{4} = \cancel{2}x + 20$$
$$\underline{-2x + \cancel{4} \quad -\cancel{2}x + 4}$$
$$\frac{\cancel{12}x}{\cancel{12}} = \frac{24}{12}$$
$$x = 2$$

## EXERCISE 2-6

1. The accompanying table shows the statements and reasons that lead to the triangles being congruent because of the side-angle-side postulate. If two triangles are congruent, then their corresponding angles are congruent.

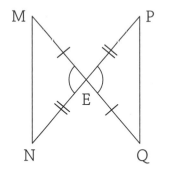

| Statement | Reason |
|---|---|
| 1. $\overline{PN}$ and $\overline{MQ}$ bisect each other at E. | 1. Given |
| 2. $\overline{ME} \cong \overline{QE}$, $\overline{NE} \cong \overline{PE}$ | 2. A bisector divides a segment into two congruent segments. |
| 3. $\angle MEN$ and $\angle QEP$ are vertical angles. | 3. Intersecting lines form vertical angles. |
| 4. $\angle MEN \cong \angle QEP$ | 4. Vertical angles are congruent. |
| 5. $\triangle MEN \cong \triangle PEQ$ | 5. SAS $\cong$ SAS |
| 6. $\angle M \cong \angle Q$ | 6. CPCTC |

2. The accompanying table shows the statements and reasons that lead to the triangles being congruent because of the angle-side-angle postulate.

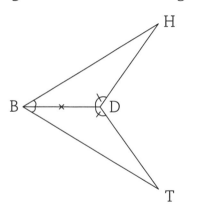

| Statement | Reason |
|---|---|
| 1. $\overline{BD}$ bisects $\angle HBT$. | 1. Given |
| 2. $\angle HBD \cong \angle TBD$ | 2. A bisector divides an angle into two congruent angles. |
| 3. $\angle HDB \cong \angle TDB$ | 3. Given |
| 4. $\overline{BD} \cong \overline{BD}$ | 4. Reflexive property |
| 5. $\triangle BHD \cong \triangle BTD$ | 5. ASA $\cong$ ASA |

3. The accompanying table shows the statements and reasons that lead to the triangles being congruent because of the side-angle-side postulate.

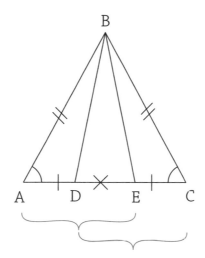

| Statement | Reason |
|---|---|
| 1. $\overline{AE} \cong \overline{CD}$ | 1. Given |
| 2. $\overline{DE} \cong \overline{DE}$ | 2. Reflexive property |
| 3. $\overline{AD} \cong \overline{CE}$ | 3. Subtraction postulate |
| 4. $\triangle ABC$ is isosceles with vertex $B$. | 4. Given |
| 5. $\overline{AB} \cong \overline{CB}$ | 5. The legs of an isosceles triangle are congruent. |
| 6. $\angle A \cong \angle C$ | 6. The base angles of an isosceles triangle are congruent. |
| 7. $\triangle ABD \cong \triangle CBE$ | 7. SAS $\cong$ SAS |

# Classifying Triangles

## EXERCISE 3-1

1. **$AC = 10$**   The legs of an isosceles triangle are equal. The perimeter of a triangle is equal to the sum of the three sides of the triangle. $x + 6$, $x + 6$, and $4x - 30$ represent the lengths of the sides of the triangle; therefore

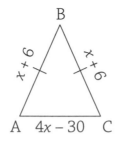

$$x + 6 + x + 6 + 4x - 30 = 42$$
$$6x - 18 = 42$$
$$\underline{+18 \quad +18}$$
$$\frac{6x}{6} = \frac{60}{6}$$
$$x = 10$$
$$AC = 4x - 30$$
$$AC = 4(10) - 30$$
$$AC = 10$$

2. **Right triangle**   A triangle is a right triangle if it contains a $90°$ angle. The sum of the three angles in a triangle must equal $180°$. $m\angle S = 3x + 9$, $m\angle I = 8x + 2$, and $m\angle T = 5x - 7$; therefore

$$m\angle S + m\angle I + m\angle T = 180°$$
$$3x + 9 + 8x + 2 + 5x - 7 = 180°$$
$$16x + \cancel{4} = 180$$
$$\underline{-\cancel{4} \quad -4}$$
$$\frac{16x}{16} = \frac{176}{16}$$
$$x = 11$$

| | | |
|---|---|---|
| $m\angle S = 3x + 9$ | $m\angle I = 8x + 2$ | $m\angle T = 5x - 7$ |
| $m\angle S = 3(11) + 9$ | $m\angle I = 8(11) + 2$ | $m\angle T = 5(11) - 7$ |
| $m\angle S = 42°$ | $m\angle I = 90°$ | $m\angle T = 48°$ |

3. **Obtuse triangle**   A triangle is an obtuse triangle if it contains an angle between 90° and 180°. The sum of the three angles in a triangle must equal 180°. $m\angle R = 2x + 4$, $m\angle E = 3x + 5$ and $m\angle D = 13x + 9$; therefore

$$m\angle R + m\angle E + m\angle D = 180°$$
$$2x + 4 + 3x + 5 + 13x + 9 = 180°$$
$$18x + \cancel{18} = 180$$
$$\underline{-\cancel{18} \quad -18}$$
$$\frac{18x}{18} = \frac{162}{18}$$
$$x = 9$$

| | | |
|---|---|---|
| $m\angle R = 2x + 4$ | $m\angle E = 3x + 5$ | $m\angle D = 13x + 9$ |
| $m\angle R = 2(9) + 4$ | $m\angle E = 3(9) + 5$ | $m\angle D = 13(9) + 9$ |
| $m\angle R = 22°$ | $m\angle E = 32°$ | $m\angle D = 126°$ |

4. **$x = 8$**  All three angles of an equilateral triangle are 60°. $m\angle C = 7x + 4$; therefore

$$7x + \cancel{4} = 60$$
$$\underline{-\cancel{4} \quad -4}$$
$$\frac{\cancel{7}x}{\cancel{7}} = \frac{56}{7}$$
$$x = 8$$

5. **Isosceles triangle**  An isosceles triangle has two equal angles. The sum of the three angles in a triangle must equal 180. $m\angle C = 6x + 5, m\angle A = 15x + 8,$ and $m\angle T = 7x - 1$; therefore

$$m\angle C + m\angle A + m\angle T = 180°$$
$$6x + 5 + 15x + 8 + 7x - 1 = 180°$$
$$28x + \cancel{12} = 180$$
$$\underline{-\cancel{12} \quad -12}$$
$$\frac{2\cancel{8}x}{2\cancel{8}} = \frac{168}{28}$$
$$x = 6$$

$m\angle C = 6x + 5 \qquad m\angle A = 15x + 8 \qquad m\angle T = 7x - 1$
$m\angle C = 6(6) + 5 \quad m\angle A = 15(6) + 8 \quad m\angle T = 7(6) - 1$
$m\angle C = 41 \qquad\quad m\angle A = 98 \qquad\qquad m\angle T = 41$

6. **$\angle C$**  The largest angle of a triangle is found opposite the largest side of the triangle. $AB = 12, BC = 10,$ and $AC = 8$; therefore

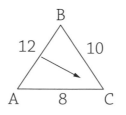

7. **∠M**   The smallest angle of a triangle is found opposite the smallest side of the triangle. $MI = 16$, $IT = 9$, and $MT = 20$; therefore

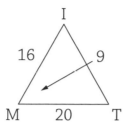

8. **$\overline{ST}, \overline{RT}, \overline{RS}$**   The smallest side of a triangle is found opposite the smallest angle of the triangle. The largest side of a triangle is found opposite the largest angle of the triangle. $m\angle R = 35°$, $m\angle S = 70°$, and $m\angle T = 75°$; therefore

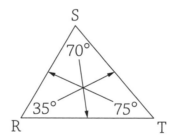

9. **15**   The sum of the two smaller sides of a triangle must be larger than the largest side of the triangle. $AB = 6$ and $BC = 10$; therefore

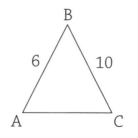

$$6 + 10 > AC \qquad \cancel{6} + AC > 10$$

$$16 > AC \qquad \frac{-\cancel{6} \qquad \quad -6}{AC > 4}$$

$$4 < AC < 16$$

**10. 4**   The sum of the two smaller sides of a triangle must be larger than the largest side of the triangle. $MN = 12$ and $NO = 9$; therefore

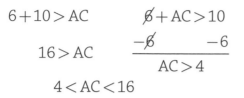

$$12 + 9 > MO \qquad \cancel{9} + MO > 12$$

$$21 > MO \qquad \frac{-\cancel{9} \qquad \quad -9}{MO > 3}$$

$$3 < MO < 21$$

## EXERCISE 3-2

**1. $x = 20$**   The sum of the three angles in a triangle must equal 180°. $m\angle A = x, m\angle B = 3x + 15,$ and $m\angle C = 4x + 5$; therefore

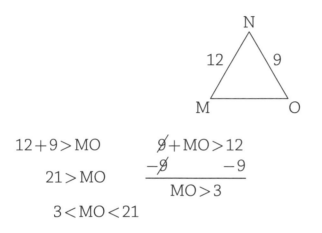

2. **$\angle E = 42°$**   The sum of the three angles in a triangle must equal 180°. $m\angle D = 9x, m\angle E = 4x - 6,$ and $m\angle F = 3x - 6$; therefore

$$m\angle E + m\angle D + m\angle F = 180°$$
$$4x - 6 + 9x + 3x - 6 = 180$$
$$16x - 12 = 180$$
$$\underline{+12 \quad +12}$$
$$\frac{16x}{16} = \frac{192}{16}$$
$$x = 12$$
$$m\angle E = 4x - 6$$
$$m\angle E = 4(12) - 6$$
$$m\angle E = 42°$$

3. **$m\angle Q = 30°$**   The exterior angle of a triangle is equal to the sum of the two nonadjacent interior angles. $\angle QRS$ is the exterior angle. $\angle P$ and $\angle Q$ are the two nonadjacent interior angles; therefore

$$m\angle P + m\angle Q = m\angle QRS$$
$$80 + m\angle Q = 110°$$
$$\underline{-80 \qquad\quad -80}$$
$$m\angle Q = 30°$$

4. **$\angle ADE = 136°$**   The exterior angle of a triangle is equal to the sum of the two nonadjacent interior angles. $\angle ADE$ is the exterior angle. $\angle M$ and $\angle A$ are the two nonadjacent interior angles; therefore

$$m\angle M + m\angle A = m\angle ADE$$
$$4x + 9 + 5x + 1 = 10x - 4$$
$$9x + 10 = 10x - 4$$
$$\underline{-9x + 4 \quad -9x + 4}$$
$$14 = x$$
$$\angle ADE - 10x - 4$$
$$\angle ADE = 10(14) - 4$$
$$\angle ADE = 136°$$

5. $\angle O$   The sum of the three angles in a triangle must equal 180°.
$m\angle D = 4x - 6$, $m\angle O = 6x - 1$, and $m\angle G = 5x + 7$; therefore

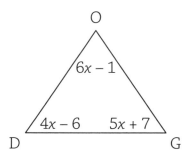

$$m\angle D + m\angle O + m\angle G = 180°$$
$$4x - 6 + 6x - 1 + 5x + 7 = 180$$
$$\frac{15x}{15} = \frac{180}{15}$$
$$x = 12$$

$$m\angle D = 4x - 6 \qquad m\angle O = 6x - 1 \qquad m\angle G = 5x + 7$$
$$m\angle D = 4(12) - 6 \quad m\angle O = 6(12) - 1 \quad m\angle G = 5(12) + 7$$
$$m\angle D = 42° \qquad m\angle O = 71° \qquad m\angle G = 67°$$

6. **∠PQS ≅ ∠RQS**   We arrived at this conclusion because we were given that $\overline{QS}$ bisected ∠PQR, and an angle bisector divides an angle into two congruent angles.

7. **$\overline{SL} \cong \overline{LT}$**   We arrived at this conclusion because we were given that $\overline{RL}$ is a median, and a median divides a segment into two congruent segments.

8. **∠ATL and ∠ATS are right angles, ∠ATL ≅ ∠ATS**   We arrived at this conclusion because we were given that $\overline{AT}$ is an altitude and an altitude forms right angles. All right angles are congruent.

9. **∠DAC = 65°**   We arrived at this conclusion because we were given that $\overline{AD}$ is an altitude, and an altitude forms right angles. The sum of the three angles in a triangle must equal 180°. $m\angle C = 25°$ and $m\angle ADC = 90°$; therefore

$$m\angle ADC + m\angle C + m\angle DAC = 180°$$
$$90 + 25 + m\angle DAC = 180°$$
$$\cancel{115} + m\angle DAC = 180$$
$$\underline{-\cancel{115} \qquad\qquad -115}$$
$$m\angle DAC = 65°$$

10. **24**   In an isosceles triangle, the two legs are equal; the altitude, median, and angle bisector are all the same segment. $DE = x + 6$, $DF = 3x$, and $EG = x + 4$; therefore

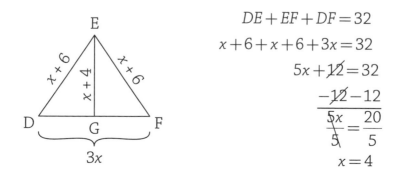

$$DE + EF + DF = 32$$
$$x + 6 + x + 6 + 3x = 32$$
$$5x + \cancel{12} = 32$$
$$\underline{-\cancel{12} \quad -12}$$
$$\frac{\cancel{5}x}{\cancel{5}} = \frac{20}{5}$$
$$x = 4$$

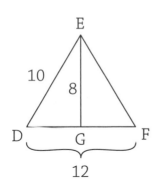

$$DE = x + 6 \qquad\qquad EF = DE$$
$$DE = 4 + 6 \qquad\qquad EF = 10$$
$$DE = 10$$

$$DF = 3x \qquad\qquad EG = x + 4$$
$$DF = 3(4) \qquad\qquad EG = 4 + 4$$
$$DF = 12 \qquad\qquad EG = 8$$

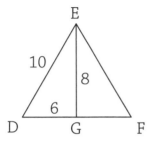

$$DG = \frac{DF}{2}$$
$$DG = \frac{12}{2}$$
$$DG = 6$$

Perimeter of $\triangle DEG$:

$6 + 8 + 10$

$= 24$

# Centers of a Triangle

## EXERCISE 4-1

1. **Centroid**   The centroid is found where the medians of the triangle intersect.

2. **Right triangle**    The orthocenter is on the vertex of a triangle when the triangle is right.

3. **False**    The Euler line does not contain the incenter. It contains the centroid, circumcenter, and orthocenter.

4. **8**    The Euler line contains the centroid, circumcenter, and orthocenter. We were given that the orthocenter is $(6,y)$; therefore,

$$y = \frac{2}{3}x + 4$$

$$y = \frac{2}{3}(6) + 4$$

$$y = 8$$

5. **Incenter**    The incenter is found where the angle bisectors of the triangle intersect. A bisector divides an angle into two congruent angles.

## EXERCISE 4-2

1. **20°**    The incenter is found where the angle bisectors of the triangle intersect. A bisector divides an angle into two congruent angles. Point $C$ represents the incenter of the triangle, and $m\angle CSI = 20°$; therefore

$$m\angle CSI \cong m\angle CSO$$

$$m\angle CSO = 20°$$

2. **50°**    We were given $\angle O \cong \angle S$.
$$m\angle S = m\angle O = 40°$$

The sum of the three angles of a triangle is $180°$.

$$m\angle SIO + m\angle IOS + m\angle OSI = 180$$
$$m\angle SIO + 40 + 40 = 180$$
$$m\angle SIO + 80 = 180$$
$$\underline{-80 \quad -80}$$
$$m\angle SIO = 100°$$

The incenter is found where the angle bisectors of the triangle intersect. A bisector divides an angle into two congruent angles.

$$m\angle CIS = \frac{100}{2} = 50°$$

3. **110°** The sum of the three angles of a triangle is 180°. The previous two problems found that $m\angle CIS = 50°$ and $m\angle ISC = 20°$; therefore

$$m\angle CIS + m\angle ISC + m\angle SCI = 180$$
$$50 + 20 + m\angle SCI = 180$$
$$\cancel{70} + m\angle SCI = 180$$
$$\underline{-\cancel{70} \qquad\qquad -70}$$
$$m\angle SCI = 110°$$

## EXERCISE 4-3

1. **34°** The incenter is found where the angle bisectors of the triangle intersect. A bisector divides an angle into two congruent angles. We were given $m\angle PCN = 34°$; therefore

$$m\angle PCN \cong m\angle PCI$$
$$m\angle PCI = 34°$$

2. **53** The incenter is found where the angle bisectors of the triangle intersect. A bisector divides an angle into two congruent angles. We were given $m\angle PIN = 48°$ and $m\angle NIC = 2x - 10$; therefore

$$m\angle NIC = 2(m\angle PIN)$$
$$2x - 10 = 2(48)$$
$$2x - \cancel{10} = 96$$
$$\underline{+\cancel{10} \; +10}$$
$$\frac{\cancel{2}x}{\cancel{2}} = \frac{106}{2}$$
$$x = 53$$

**EXERCISE 4-4**

1. **8.2**  The centroid is found where the medians of the triangle intersect. A median divides a segment into two congruent segments.

2. **11.25**  The centroid is found where the medians of the triangle intersect. A median divides a segment into two congruent segments.

3. **8.5**  The centroid of a triangle divides each median of the triangle into segments with a 2:1 ratio. We were given $CA = 17$; therefore

$$\frac{CA}{MA} = \frac{2}{1}$$
$$\frac{17}{MA} = \frac{2}{1}$$
$$\frac{2MA}{2} = \frac{17}{2}$$
$$MA = 8.5$$

4. **36**  The centroid of a triangle divides each median of the triangle into segments with a 2:1 ratio. We were given $CM = 54$; therefore

$$\text{let } x = MA$$
$$\text{let } 2x = CA$$
$$2x + x = 54$$
$$\frac{3x}{3} = \frac{54}{3}$$
$$x = 18$$
$$CA = 2(18) = 36$$

5. **8**  The centroid of a triangle divides each median of the triangle into segments with a 2:1 ratio. We were given $DA = 14$ and $NA = 4x - 4$; therefore

$$\frac{NA}{DA} = \frac{2}{1}$$

$$\frac{4x-4}{14} = \frac{2}{1}$$

$$4x - \cancel{4} = 28$$

$$\underline{+\cancel{4} \quad +4}$$

$$\frac{\cancel{4}x}{\cancel{4}} = \frac{32}{4}$$

$$x = 8$$

6. **27**   The centroid of a triangle divides each median of the triangle into segments with a 2:1 ratio. We were given $DN = 81$; therefore

$$DA = x$$

$$NA = 2x$$

$$x + 2x = 81$$

$$\frac{\cancel{3}x}{\cancel{3}} = \frac{81}{3}$$

$$x = 27$$

$$DA = 27$$

## EXERCISE 4-5

1. **90°**   The orthocenter is found where the altitudes of the triangle intersect. The altitude of a triangle is a segment drawn from the vertex of the triangle perpendicular to the opposite side of the triangle forming congruent right angles.

2. **20**   We were given that point $P$ represents the orthocenter, which means that the altitudes of the triangle intersect at point $P$. The altitude of a triangle is a segment drawn from the vertex of the triangle perpendicular to the opposite side of the triangle to form congruent right angles. We were given that $m\angle RST = 3x + 30$; therefore

$$m\angle RST = 90°$$
$$3x + 30 = 90$$
$$\underline{-30 \quad -30}$$
$$\frac{3x}{3} = \frac{60}{3}$$
$$x = 20$$

# Similarity

## EXERCISE 5-1

1. **63°**    We were given that $\triangle DEF$ is the image of $\triangle ABC$ after it is dilated by a scale factor of 2. Because a dilation creates two similar triangles, $\triangle DEF$ must be similar to $\triangle ABC$. The corresponding angles of similar triangles are congruent. $\angle A$ corresponds to $\angle D$, $\angle B$ corresponds to $\angle E$, and $\angle C$ corresponds to $\angle F$. The sum of all the angles of a triangle equals 180°. We were given $m\angle A = 50°$ and $m\angle B = 67°$; therefore

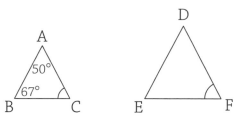

$$m\angle C = m\angle F$$

$$m\angle A + m\angle B + m\angle C = 180°$$
$$50° + 67° + m\angle C = 180°$$
$$\cancel{117°} + m\angle C = 180°$$
$$\underline{-\cancel{117°} \qquad\qquad -117°}$$
$$m\angle C = 63°$$
$$m\angle F = 63°$$

2. **20.25 units²** The ratio of the area of similar triangles is equal to the square of the ratio of the corresponding sides of the similar triangles. We were given $AC = 4$ units, $DF = 6$ units, and the area of $\triangle ABC = 9$ units²; therefore

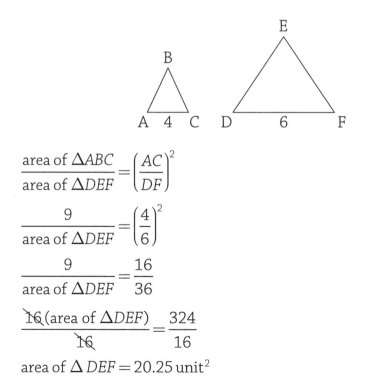

$$\frac{\text{area of } \triangle ABC}{\text{area of } \triangle DEF} = \left(\frac{AC}{DF}\right)^2$$

$$\frac{9}{\text{area of } \triangle DEF} = \left(\frac{4}{6}\right)^2$$

$$\frac{9}{\text{area of } \triangle DEF} = \frac{16}{36}$$

$$\frac{\cancel{16}\,(\text{area of } \triangle DEF)}{\cancel{16}} = \frac{324}{16}$$

$$\text{area of } \triangle DEF = 20.25 \text{ unit}^2$$

3. **10**   When an altitude is drawn in a right triangle from the right angle to the hypotenuse, it forms similar right triangles. The following proportion can be used to solve for the missing leg of the right triangle:

$$\frac{\text{Projected segment}}{\text{Leg}} = \frac{\text{Leg}}{\text{Hypotenuse}}$$

We were given $AD = 5$ and $DC = 15$; therefore

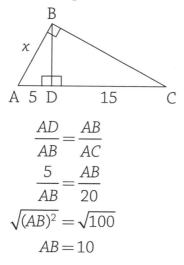

$$\frac{AD}{AB} = \frac{AB}{AC}$$
$$\frac{5}{AB} = \frac{AB}{20}$$
$$\sqrt{(AB)^2} = \sqrt{100}$$
$$AB = 10$$

4. **21**   If two triangles are similar, the largest side of one triangle corresponds with the largest side of the other triangle, which means $\overline{AC}$ corresponds to $\overline{DF}$. We were given the perimeter of $\triangle DEF$ to be 51 and can add the sides of $\triangle ABC$ to find that the perimeter of $\triangle ABC = 17$. The ratio of the perimeters of similar triangles is equal to the ratio of the corresponding sides of the triangle; therefore

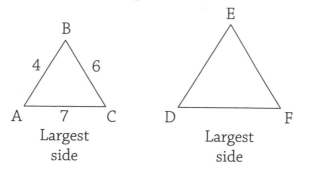

Perimeter of $\triangle ABC = 17$

$$\frac{\text{Perimeter of } \triangle ABC}{\text{Perimeter of } \triangle DEF} = \frac{AC}{DF}$$

$$\frac{17}{51} = \frac{7}{DF}$$

$$\frac{\cancel{17}(DF)}{\cancel{17}} = \frac{357}{17}$$

$$DF = 21$$

5. **No**

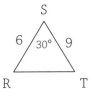   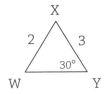

$$\frac{RS}{ST} = \frac{6}{9} = \frac{2}{3}$$

$$\frac{WX}{XY} = \frac{2}{3}$$

Even though the two corresponding sides are in proportion, the congruent angle is not included between the two sides in $\triangle WXY$. For these triangles to be similar using the SAS similarity method, $m\angle X$ would have to equal 30°.

### EXERCISE 5-2

1. **12.6**   When triangles are similar, their corresponding sides are in proportion. We were given $AB = 4.2$, $AC = 6.4$, and $DF = 19.2$; therefore

$$\frac{AB}{AC} = \frac{DE}{DF}$$

$$\frac{4.2}{6.4} = \frac{DE}{19.2}$$

$$\frac{6.4DE}{6.4} = \frac{80.64}{6.4}$$

$$DE = 12.6$$

2. **3.5**    $m\angle DEC = 75°$ because it forms a linear pair with $\angle FEC$ and angles that form a linear pair are supplementary. $\angle BCA \cong \angle DCE$ because vertical angles are congruent. This gives us enough information to say $\triangle ABC \sim \triangle EDC$ because of the AA similarity theorem.

The corresponding sides of similar triangles are in proportion. $\overline{BC}$ corresponds to $\overline{DC}$, and $\overline{AB}$ corresponds to $\overline{DE}$; therefore

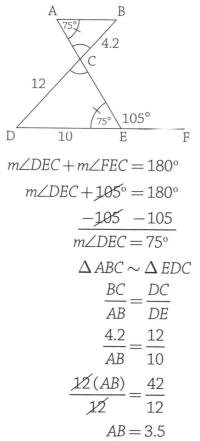

$$m\angle DEC + m\angle FEC = 180°$$

$$m\angle DEC + 105° = 180°$$

$$\frac{-105 \quad -105}{m\angle DEC = 75°}$$

$$\triangle ABC \sim \triangle EDC$$

$$\frac{BC}{AB} = \frac{DC}{DE}$$

$$\frac{4.2}{AB} = \frac{12}{10}$$

$$\frac{12(AB)}{12} = \frac{42}{12}$$

$$AB = 3.5$$

3. **15**  The ratio of the perimeter of similar triangles is equal to the ratio of the corresponding sides of the similar triangles. We were given the perimeter of $\triangle DOG$ to be 30, the perimeter of $\triangle CAT$ to be 45, and $DO = 10$; therefore

$$\frac{DO}{CA} = \frac{\text{Perimeter of } \triangle DOG}{\text{Perimeter of } \triangle CAT}$$

$$\frac{10}{CA} = \frac{30}{45}$$

$$\frac{30(CA)}{30} = \frac{450}{30}$$

$$CA = 15$$

4. **12**  When a line is drawn in a triangle parallel to a side, it creates similar triangles. $\overline{MP} \parallel \overline{NO}$, which means $\triangle LPM \sim \triangle LON$. The corresponding sides of similar triangles are in proportion. $\overline{LP}$ corresponds to $\overline{LO}$, and $\overline{MP}$ corresponds to $\overline{NO}$; therefore

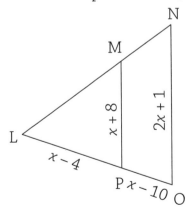

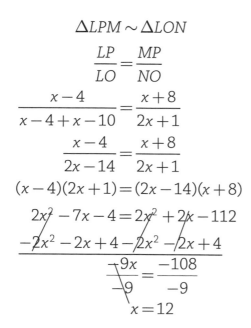

$$\triangle LPM \sim \triangle LON$$

$$\frac{LP}{LO} = \frac{MP}{NO}$$

$$\frac{x-4}{x-4+x-10} = \frac{x+8}{2x+1}$$

$$\frac{x-4}{2x-14} = \frac{x+8}{2x+1}$$

$$(x-4)(2x+1) = (2x-14)(x+8)$$

$$2x^2 - 7x - 4 = 2x^2 + 2x - 112$$

$$-2x^2 - 2x + 4 - 2x^2 - 2x + 4$$

$$\frac{-9x}{-9} = \frac{-108}{-9}$$

$$x = 12$$

5. **6**  When an altitude is drawn in a right triangle from the right angle to the hypotenuse, it forms similar right triangles. The following proportion can be used to solve for the altitude of the right triangle:

$$\frac{\text{Segment}}{\text{Altitude}} = \frac{\text{Altitude}}{\text{Remaining segment}}$$

We were given $LO = 2$ and $LN = 20$, which means $ON = 18$; therefore

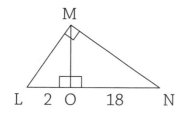

$$\frac{LO}{MO} = \frac{MO}{ON}$$

$$\frac{2}{MO} = \frac{MO}{18}$$

$$\sqrt{(MO)^2} = \sqrt{36}$$

$$MO - 6$$

6. **20 miles**

let $x$ = distance from Jonathan's house to Andrew's house

let $x - 4$ = distance from Jonathan's house to Zachary's house

let $2x$ = distance from Jonathan's house to Brandon's house

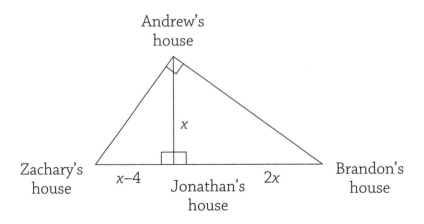

$$\frac{\text{Segment 1}}{\text{altitude}} = \frac{\text{altitude}}{\text{Segment 2}}$$

$$\frac{x-4}{x} = \frac{x}{2x}$$

$$x^2 = 2x(x-4)$$

$$\cancel{x^2} = 2x^2 - 8x$$

$$\frac{-\cancel{x^2} \quad -x^2}{0 = x^2 - 8x}$$

$$0 = x(x-8)$$

$$x = 0 \, \text{\o} \, x = 8$$

Distance from Zachary's house to Brandon's house
$$= x - 4 + 2x$$
$$= 8 - 4 + 2(8)$$
$$= 20 \text{ miles}$$

**7. 5** When an altitude is drawn in a right triangle from the right angle to the hypotenuse, it forms similar right triangles. The following proportion can be used to solve for the missing variable:

$$\frac{\text{Projected segment}}{\text{Leg}} = \frac{\text{Leg}}{\text{Hypotenuse}}$$

We were given $BC = 2x + 2$, $AD = 3x - 8$, and $DC = x + 4$; therefore

$$\frac{DC}{BC} = \frac{BC}{AC}$$

$$\frac{x+4}{2x+2} = \frac{2x+2}{x+4+3x-8}$$

$$\frac{x+4}{2x+2} = \frac{2x+2}{4x-4}$$

$$(2x+2)(2x+2) = (x+4)(4x-4)$$

$$4x^2 + 8x + 4 = 4x^2 + 12x - 16$$

$$-4x^2 - 12x - 4 - 4x^2 - 12x - 4$$

$$\frac{-4x}{-4} = \frac{-20}{-4}$$

$$x = 5$$

## EXERCISE 5-3

1. The accompanying table shows the statements and reasons for the markings that can be placed in the given diagram. From these markings, we can see that the triangles are similar because of the angle-angle similarity postulate.

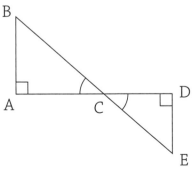

| Statement | Reason |
|---|---|
| 1. $\overline{BE}$ intersects $\overline{AD}$ at C. | 1. Given |
| 2. $\angle BCA \cong \angle ECD$ | 2. Intersecting lines form congruent vertical angles. |
| 3. $\overline{BA} \perp \overline{AD}, \overline{ED} \perp \overline{AD}$ | 3. Given |
| 4. $\angle BAC$ and $\angle EDC$ are right angles. | 4. Perpendicular lines form right angles. |
| 5. $\angle BAC \cong \angle EDC$ | 5. All right angles are congruent. |
| 6. $\triangle BAC \sim \triangle EDC$ | 6. AA $\cong$ AA |

**2.** The accompanying table shows the statements and reasons for the markings that can be placed in the given diagram. From these markings, we can see (a) that the triangles are similar because of the angle-angle similarity postulate and (b) that the sides are in proportion because the corresponding sides of similar triangles are in proportion.

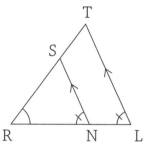

| Statement | Reason |
|---|---|
| 1. $\triangle RTL, \overline{SN} \parallel \overline{TL}$ | 1. Given |
| 2. $\angle SNR \cong \angle TLR$ | 2. When parallel lines are cut by a transversal, they form congruent corresponding angles. |
| 3. $\angle R \cong \angle R$ | 3. Reflexive property |
| 4. $\triangle RNS \sim \triangle RLT$ | 4. AA $\cong$ AA |
| 5. $\dfrac{RN}{RL} = \dfrac{SN}{TL}$ | 5. Corresponding sides of similar triangles are in proportion. |

3. The accompanying table shows the statements and reasons for the markings that can be placed in the given diagram. From these markings, we can see that the triangles are similar because of the angle-angle similarity postulate and that the sides are in proportion because the corresponding sides of similar triangles are in proportion. In a proportion, the product of the means equals the product of the extremes.

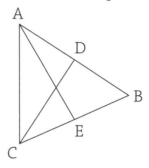

| Statement | Reason |
|---|---|
| 1. $\triangle ABC$ is isosceles with vertex at $\angle C$, $\overline{AE} \perp \overline{CB}$, $\overline{CD} \perp \overline{AB}$ | 1. Given |
| 2. $\angle CAB \cong \angle CBA$ | 2. The base angles of an isosceles triangle are congruent. |
| 3. $\angle CDA$ and $\angle AEB$ are right angles. | 3. Perpendicular lines form right angles. |
| 4. $\angle CDA \cong \angle AEB$ | 4. If two angles are right angles, they are congruent. |
| 5. $\triangle CDA \sim \triangle AEB$ | 5. AA $\cong$ AA |
| 6. $\dfrac{AD}{BE} = \dfrac{CD}{AE}$ | 6. Corresponding sides of similar triangles are in proportion. |
| 7. $BE \times CD = AD \times AE$ | 7. In a proportion, the product of the means equals the product of the extremes. |

# Getting to Know Right Triangles

**EXERCISE 6-1**

1. **8**

$$\text{leg}^2 + \text{leg}^2 = \text{hypotenuse}^2$$
$$x^2 + 15^2 = 17^2$$
$$x^2 + 225 = 289$$
$$\underline{-225 \quad -225}$$
$$\sqrt{x^2} = \sqrt{64}$$
$$x = 8$$

2. **No,** it is not a right triangle because it does not satisfy the Pythagorean theorem.

$$\text{leg}^2 + \text{leg}^2 \overset{?}{=} \text{hypotenuse}^2$$
$$17^2 + 15^2 \overset{?}{=} 25^2$$
$$289 + 225 \overset{?}{=} 625$$
$$514 \neq 625$$

3. $AB = 3\sqrt{2}$

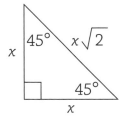

$$x = 3$$
$$AB = x\sqrt{2}$$
$$AB = 3\sqrt{2}$$

4. $AC = 10$

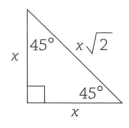

$$\frac{x\sqrt{2}}{\sqrt{2}} = \frac{10\sqrt{2}}{\sqrt{2}}$$
$$x = 10$$
$$AC = 10$$

5. $BC = \dfrac{15\sqrt{2}}{2}$

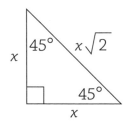

$$\frac{x\sqrt{2}}{\sqrt{2}} = \frac{15}{\sqrt{2}}\left(\frac{\sqrt{2}}{\sqrt{2}}\right)$$
$$x = \frac{15\sqrt{2}}{2}$$
$$BC = \frac{15\sqrt{2}}{2}$$

6. $AB = 8$

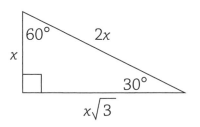

$$x = 4$$
$$AB = 2x$$
$$AB = 2(4)$$
$$AB = 8$$

7. $BC = 6\sqrt{3}$

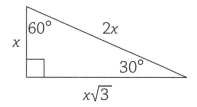

$$x = 6$$
$$BC = x\sqrt{3}$$
$$BC = 6\sqrt{3}$$

8. $AC = 7, BC = 7\sqrt{3}$

$$\frac{\cancel{2}x}{\cancel{2}} = \frac{14}{2}$$
$$x = 7$$
$$AC = x$$
$$AC = 7$$
$$BC = x\sqrt{3}$$
$$BC = 7\sqrt{3}$$

9. $AC = \dfrac{8\sqrt{3}}{3}$, $AB = \dfrac{16\sqrt{3}}{3}$

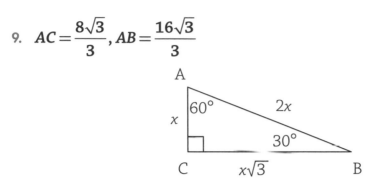

$$\frac{x\sqrt{3}}{\sqrt{3}} = \frac{8}{\sqrt{3}}\left(\frac{\sqrt{3}}{\sqrt{3}}\right)$$

$$x = \frac{8\sqrt{3}}{3}$$

$AC = x \qquad AB = 2x$

$AC = \dfrac{8\sqrt{3}}{3} \qquad AB = 2\left(\dfrac{8\sqrt{3}}{3}\right)$

$$AB = \frac{16\sqrt{3}}{3}$$

10. $x = 29.2$

$$\sin\theta = \frac{\text{opposite}}{\text{hypotenuse}}$$

$$\frac{\sin 20°}{1} = \frac{10}{x}$$

$$\frac{x\ \sin 20°}{\sin 20°} = \frac{10}{\sin 20°}$$

$$x = 29.2$$

11. **x = 7.62**

$$\cos\theta = \frac{\text{adjacent}}{\text{hypotenuse}}$$

$$\frac{\cos 57°}{1} = \frac{x}{14}$$

$$x = 14\cos 57°$$

$$x = 7.62$$

12. **x = 52.1°**

$$\tan\theta = \frac{\text{opposite}}{\text{adjacent}}$$

$$\tan x = \frac{9}{7}$$

$$x = \tan^{-1}\left(\frac{9}{7}\right)$$

$$x = 52.1°$$

13. **x = 31.23°**

$$\sin\theta = \frac{\text{opposite}}{\text{hypotenuse}}$$

$$\sin x = \frac{8.4}{16.2}$$

$$x = \sin^{-1}\left(\frac{8.4}{16.2}\right)$$

$$x = 31.23°$$

14. **99 ft**

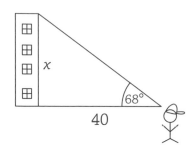

$$\tan\theta = \frac{\text{opposite}}{\text{hypotenuse}}$$

$$\frac{\tan 68°}{1} = \frac{x}{40}$$

$$x = 40 \tan 68°$$

$$x = 99 \,\text{ft}$$

15. **70.2°**   The angle of depression is equal to the angle of elevation. The entire height of the triangle is the height of the building plus Sherry's height.

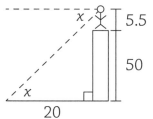

$$\tan\theta = \frac{\text{opposite}}{\text{adjacent}}$$

$$\tan x = \frac{55.5}{20}$$

$$x = \tan^{-1}\left(\frac{55.5}{20}\right)$$

$$x = 70.2°$$

# Parallel Lines

## EXERCISE 7-1

1. **Yes**    If two parallel lines are cut by a transversal, their alternate interior angles are congruent.

2. **Yes**    If two parallel lines are cut by a transversal, their alternate exterior angles are congruent.

3. **No**    These angles form a linear pair, which means they are supplementary.

4. **Yes**    If two parallel lines are cut by a transversal, their corresponding angles are congruent.

5. **No**    If two parallel lines are cut by a transversal, the same side exterior angles are supplementary.

6. **No**    If two parallel lines are cut by a transversal, the same side interior angles are supplementary.

7. **Yes**    Intersecting lines form vertical angles that are congruent.

## EXERCISE 7-2

1. **108°**    If two parallel lines are cut by a transversal, their alternate interior angles are congruent. We were given $m\angle AGH = 108°$; therefore, $m\angle AGH = m\angle DHG = 108°$.

2. **20**    If two parallel lines are cut by a transversal, the same side interior angles are supplementary. We were given $m\angle CHG = 3x + 2$ and $m\angle AGH = 5x + 18$; therefore

$$m\angle CHG + m\angle AGH = 180°$$
$$3x + 2 + 5x + 18 = 180$$
$$8x + 20 = 180$$
$$\frac{-20 \quad -20}{8} \quad x = \frac{160}{8}$$
$$x = 20$$

3. **38°** If two parallel lines are cut by a transversal, their alternate interior angles are congruent. We were given $m\angle BGH = 3x + 20$ and $m\angle CHG = x + 32$; therefore

$$m\angle BGH = m\angle CHG$$
$$3x + 20 = x + 32$$
$$\frac{-x - 20 \quad -x - 20}{2x} = \frac{12}{2}$$
$$x = 6$$
$$m\angle BGH = 3x + 20$$
$$m\angle BGH = 3(6) + 20$$
$$m\angle BGH = 38°$$

4. **136°** If two parallel lines are cut by a transversal, their corresponding angles are congruent. We were given $m\angle EGB = 6x - 14$ and $m\angle GHD = 3x + 61$; therefore

$$m\angle EGB = m\angle GHD$$
$$6x - 14 = 3x + 61$$
$$\frac{-3x + 14 \quad -3x + 14}{3x} = \frac{75}{3}$$
$$x = 25$$

$m\angle GHD = 3x + 61$

$m\angle GHD = 3(25) + 61$

$m\angle GHD = 136°$

5. **40**   Intersecting lines form congruent vertical angles which is why $m\angle GHC = m\angle DHF$. If two parallel lines are cut by a transversal, the same side interior angles, $\angle GHC$ and $\angle AGH$, are supplementary. We were given $m\angle AGH = 4x - 45$ and $m\angle DHF = 2x - 15$; therefore

$$m\angle GHC = m\angle DHF$$

$$m\angle GHC = 2x + 15$$

$$m\angle GHC + m\angle AGH = 180°$$

$$2x - 15 + 4x - 45 = 180°$$

$$6x - \cancel{60} = 180°$$

$$\underline{+\cancel{60} \quad +60}$$

$$\frac{\cancel{6}x}{\cancel{6}} = \frac{240}{6}$$

$$x = 40$$

## EXERCISE 7-3

1. **No**

   $m\angle BAC = 4(20) - 45 = 35°$

   $m\angle DCE = 2(20) + 15 = 55°$

   $\overline{AB}$ is *not* parallel to $\overline{CD}$ because the corresponding angles are *not* congruent. The angles need to be congruent for the lines to be parallel.

2. **Yes**

   $m\angle DEC = 10x - 10 = 10(3) - 10 = 20°$

   $m\angle BCA = 2x + 14 = 2(3) + 14 = 20°$

   $\overline{BC} \parallel \overline{DE}$ because the corresponding angles are congruent.

3. **72°**   If two parallel lines are cut by a transversal, their corresponding angles are congruent. We were given $m\angle A = x + 27$ and $m\angle DCE = 2x - 18$; therefore

$$m\angle A = m\angle DCE$$
$$x + 27 = 2x - \cancel{18}$$
$$\underline{-x + 18 \quad -x + \cancel{18}}$$
$$45 = x$$
$$m\angle A = x + 27$$
$$m\angle A = 45 + 27$$
$$m\angle A = 72°$$

4.

| Statement | Reason |
|---|---|
| 1. $\overline{AB} \parallel \overline{CD}$ <br> $\overline{DE} \parallel \overline{BC}$ | 1. Given |
| 2. $\angle BAC$ and $\angle DCE$, <br> $\angle DEC$ and $\angle BCA$ are corresponding angles. | 2. Parallel lines cut by a transversal form corresponding angles. |
| 3. $\angle BAC \cong \angle DCE$, <br> $\angle DEC \cong \angle BCA$ | 3. Corresponding angles are congruent. |
| 4. $C$ is the midpoint of $\overline{AE}$. | 4. Given |
| 5. $\overline{AC} \cong \overline{EC}$ | 5. A midpoint divides a segment into two congruent segments. |
| 6. $\triangle ABC \cong \triangle CDE$ | 6. ASA $\cong$ ASA |
| 7. $\angle B \cong \angle D$ | 7. If two triangles are congruent, then their corresponding angles are congruent. |

## EXERCISE 7-4

1. **43°**   If two parallel lines are cut by a transversal, their alternate interior angles are congruent. We were given $m\angle ATH = 43°$; therefore

$$m\angle MAT = m\angle ATH = 43°$$

2. **36°**    The sum of the angles of a triangle is 180°. We were given $m\angle M = 101°$ and in the previous example we found $m\angle MAT = 43°$; therefore

$$m\angle M + m\angle A + m\angle ATM = 180°$$
$$101 + 43 + m\angle ATM = 180$$
$$\cancel{144} + m\angle ATM = 180$$
$$\underline{-\cancel{144} \qquad\qquad -144}$$
$$m\angle ATM = 36°$$

## EXERCISE 7-5

| Statement | Reason |
|---|---|
| $\overline{PA} \cong \overline{LE}$, $\overline{RA} \cong \overline{ME}$ | Given |
| $\overline{PL} \cong \overline{MR}$ | Given |
| 1. $\overline{RL} \cong \overline{RL}$ | Reflexive property |
| $\overline{PR} \cong \overline{ML}$ | 2. **Subtraction postulate** |
| $\triangle APR \cong \triangle ELM$ | 3. **SSS $\cong$ SSS** |
| $\angle APR \cong \angle ELM$ | If two triangles are congruent, then their corresponding angles are congruent. |
| $\overline{PA} \parallel \overline{LE}$ | 4. **If two lines are cut by a transversal such that their corresponding angles are congruent, then the lines are parallel.** |

## EXERCISE 7-6

1. **96°** Draw an auxiliary line that is parallel to $\overline{GM}$ and $\overline{SA}$.

$m\angle SIG = 35° + 61° = 96°$

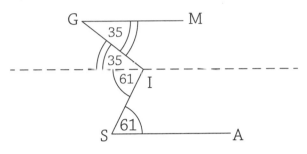

2. **112°** Draw an auxiliary line parallel to $\overline{PN}$ and $\overline{ST}$. When two parallel lines are cut by a transversal, same-side interior angles are supplementary.

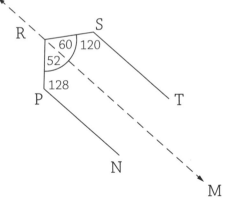

$m\angle RPN + m\angle PRM = 180°$      $m\angle RST + m\angle SRM = 180°$

$128° + m\angle PRM = 180°$      $120° + m\angle SRM = 180°$

$m\angle PRM = 52°$      $m\angle SRM = 60°$

$\downarrow$

$m\angle PRS = m\angle PRM + m\angle SRM$

$m\angle PRS = 52° + 60°$

$m\angle PRS = 112°$

# Parallelograms

## EXERCISE 8-1

1. **43**   Opposite sides of a parallelogram are congruent. We were given $BC = 22x - 1$ and $AD = 3x + 37$; therefore

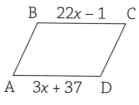

$$BC = AD$$
$$22x - 1 = 3x + 37$$
$$\underline{-3x + 1 \quad -3x + 1}$$
$$\frac{19x}{19} = \frac{38}{19}$$
$$x = 2$$
$$AD = 3x + 37$$
$$AD = 3(2) + 37$$
$$AD = 43$$

2. **$m\angle C = 37°$ and $m\angle B = 143°$**   Opposite angles of a parallelogram are congruent, and consecutive angles of a parallelogram are supplementary. We were given $m\angle A = 37°$; therefore

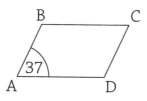

$$m\angle A = m\angle C$$
$$37° = m\angle C$$
$$m\angle A + m\angle B = 180°$$
$$\cancel{37} + m\angle B = 180$$
$$\underline{-\cancel{37} \qquad\quad -37}$$
$$m\angle B = 143°$$

## EXERCISE 8-2

1. **False**. Only the diagonals of a square and rhombus bisect the angles.

2. **False**. Only the diagonals of a rectangle, a square, and an isosceles trapezoid are congruent.

3. **True**

4. **True**

5. **True**

6. **False**. A rectangle does not have all sides congruent. Therefore, a rectangle is *not* a square.

7. **True**

8. **True**

9. **True**

10. **True**

## EXERCISE 8-3

1. **15** The diagonals of a rhombus are perpendicular, which means they create four right triangles. We were given $RM = 12$ and $SM = 9$; therefore

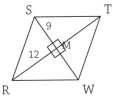

$$a^2 + b^2 = c^2$$
$$(12)^2 + (9)^2 = (RS)^2$$
$$144 + 81 = (RS)^2$$
$$\sqrt{225} = \sqrt{(RS)^2}$$
$$15 = RS$$

2. **52**    The diagonals of a rhombus are perpendicular bisectors. They divide the rhombus into four right triangles with the legs being half the length of the diagonals. We can then use the Pythagorean theorem to find the length of the hypotenuse of the triangle, which is really a side of the rhombus. The sides of a rhombus are all congruent, so the perimeter of a rhombus is the sum of the four congruent sides or four times the length of one side. We were given diagonals $RT = 24$ and $SW = 10$; therefore

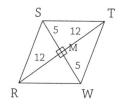

$$a^2 + b^2 = c^2$$
$$(12)^2 + (5)^2 = (ST)^2$$
$$144 + 25 = (ST)^2$$
$$\sqrt{169} = \sqrt{(ST)^2}$$
$$13 = ST$$
$$4(13) = 52$$

**3. 12**  A rhombus has four congruent sides.  The perimeter of a rhombus is the sum of the four sides of the rhombus. We were given that the perimeter of the rhombus is 100 and $WT = 2x + 1$; therefore

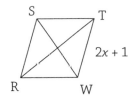

$$4(WT) = 100$$
$$4(2x + 1) = 100$$
$$8x + \cancel{4} = 100$$
$$\underline{-\cancel{4} \quad -4}$$
$$\frac{\cancel{8}x}{\cancel{8}} = \frac{96}{8}$$
$$x = 12$$

**4. 35**  The diagonals of a rhombus are perpendicular, and perpendicular lines form right angles. We were given $m\angle SMT = 3x - 15$; therefore

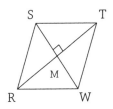

$$m\angle SMT = 90°$$
$$3x - 15 = 90$$
$$\underline{+15 \ +15}$$
$$\frac{3x}{3} = \frac{105}{3}$$
$$x = 35$$

5. **50°** The diagonals of a rhombus are perpendicular and divide the rhombus into four right triangles. We were given $m\angle MRS = 40°$ and know that $m\angle RMS = 90°$. The sum of the angles of a triangle is 180°; therefore

$$m\angle MRS + m\angle RMS + m\angle MSR = 180$$
$$40 + 90 + m\angle MSR = 180$$
$$m\angle MSR = 50°$$

## EXERCISE 8-4

1. **13** The diagonals of a rectangle are congruent. We were given that diagonals $AC = x^2 - 2x - 11$ and $BD = 2x + 1$; therefore

$$AC = BD$$
$$x^2 - 2x - 11 = \cancel{2x} + \cancel{1}$$
$$\underline{-2x - 1 \ -\cancel{2x} - \cancel{1}}$$
$$x^2 - 4x - 12 = 0$$
$$\frac{(x-6)(x+2) = 0}{x = 6 \quad x = -2} = 8$$
$$BD = 2x + 1$$
$$BD = 2(6) + 1$$
$$BD = 13$$

2. **14** A rectangle contains four right angles. We were given that $m\angle DAB = 7x - 8$; therefore

$$m\angle DAB = 90°$$
$$7x - \cancel{8} = 90$$
$$\underline{+\cancel{8} \ +8}$$
$$\frac{\cancel{7}x}{\cancel{7}} = \frac{98}{7}$$
$$x = 14$$

3. **102**   A rectangle contains four right angles. This means the diagonal divides the rectangle into two congruent right triangles with the diagonal being the hypotenuse. We were given the length of a side of the rectangle ($AB = 15$) and the diagonal ($AC = 39$). The Pythagorean theorem will solve for the other side of the rectangle. The perimeter of a rectangle is the sum of the sides, and the opposite sides of a rectangle are congruent; therefore

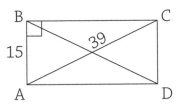

$$a^2 + b^2 = c^2$$
$$15^2 + (BC)^2 = (39)^2$$
$$225 + (BC)^2 = 1521$$
$$\underline{-225 \qquad\quad -225}$$
$$\sqrt{(BC)^2} = \sqrt{1{,}296}$$
$$BC = 36$$
$$15 + 15 + 36 + 36$$
$$= 102$$

## EXERCISE 8-5

1. **5**   The diagonals of an isosceles trapezoid are congruent. We were given $LD = 17b + 3$ and $AG = 88$; therefore

$$LD = AG$$
$$17b + 3 = 88$$
$$\underline{\quad -3 \;\; -3}$$
$$\frac{\cancel{17}b}{\cancel{17}} = \frac{85}{17}$$
$$b = 5$$

2. **140°**  The consecutive angles of a trapezoid are supplementary. We were given $m\angle LGD = 40°$; therefore

$$m\angle LGD + m\angle GLA = 180°$$
$$\cancel{40} + m\angle GLA = 180°$$
$$\underline{-\cancel{40} \qquad\qquad -40}$$
$$m\angle GLA = 140°$$

3. **80**  The bases of a trapezoid are parallel and form congruent alternate interior angles. We were given $m\angle ALD = y + 30$ and $m\angle GDL = 110$; therefore

$$m\angle ALD = m\angle GDL$$
$$y + \cancel{30} = 110$$
$$\underline{-\cancel{30} \quad -30}$$
$$y = 80$$

## EXERCISE 8-6

1. **21**  The midsegment of a trapezoid is the average of the two bases of the trapezoid. We were given $DI = 17$ and $MN = 25$; therefore

$$EA = \frac{17 + 25}{2}$$
$$EA = 21$$

2. **85**  The midsegment of a trapezoid is the average of the two bases of the trapezoid. We were given $DI = 33$ and $EA = 59$; therefore

$$\frac{33 + MN}{2} = 59$$

$$MN = 85$$

3. **2** The midsegment of a trapezoid is the average of the two bases of the trapezoid. We were given $DI = x + 4$, $MN = 3x + 6$, and $EA = x + 7$; therefore

$$EA = \frac{DI + MN}{2}$$

$$x + 7 = \frac{x + 4 + 3x + 6}{2}$$

$$\frac{x + 7}{1} = \frac{4x + 10}{2}$$

$$2x + 14 = 4x + \cancel{10}$$

$$\underline{-2x - 10 \quad -2x - \cancel{10}}$$

$$\frac{4}{2} = \frac{\cancel{2}x}{\cancel{2}}$$

$$x = 2$$

# Coordinate Geometry

**EXERCISE 9-1**

1. **10**

$$d = \sqrt{(x_2 - x_1)^2 + (y_2 - y_1)^2}$$
$$AB = \sqrt{(4 - 4)^2 + (25 - 15)^2}$$
$$AB = \sqrt{(0)^2 + (10)^2}$$
$$AB = \sqrt{100}$$
$$AB = 10$$

2. **13**

$$d = \sqrt{(x_2 - x_1)^2 + (y_2 - y_1)^2}$$
$$NP = \sqrt{(11 - (-2))^2 + (14 - 14)^2}$$
$$NP = \sqrt{(13)^2 + (0)^2}$$
$$NP = \sqrt{169}$$
$$NP = 13$$

3. **5**

$$d = \sqrt{(x_2 - x_1)^2 + (y_2 - y_1)^2}$$
$$ST = \sqrt{(6-3)^2 + (8-12)^2}$$
$$ST = \sqrt{(3)^2 + (-4)^2}$$
$$ST = \sqrt{25}$$
$$ST = 5$$

4. **$2\sqrt{17}$**

$$d = \sqrt{(x_2 - x_1)^2 + (y_2 - y_1)^2}$$
$$HK = \sqrt{(3-(-5))^2 + (1-(-1))^2}$$
$$HK = \sqrt{(8)^2 + (2)^2}$$
$$HK = \sqrt{68}$$
$$HK = \sqrt{4}\sqrt{17}$$
$$HK = 2\sqrt{17}$$

## EXERCISE 9-2

1. **Yes**

$$m\overline{BA} = \frac{y_2 - y_1}{x_2 - x_1} = \frac{3-2}{6-4} = \frac{1}{2}$$

$$m\overline{SE} = \frac{y_2 - y_1}{x_2 - x_1} = \frac{4-5}{3-5} = \frac{-1}{-2} = \frac{1}{2}$$

The segments are parallel because they have equal slopes.

2. **Yes**

$$m\overline{SB} = \frac{y_2 - y_1}{x_2 - x_1} = \frac{5-2}{5-4} = \frac{3}{1}$$

$$m\overline{AE} = \frac{y_2 - y_1}{x_2 - x_1} = \frac{4-3}{3-6} = -\frac{1}{3}$$

The diagonals are perpendicular because the slopes are negative reciprocals.

3. **Yes**

$$AS = \sqrt{(x_2 - x_1)^2 + (y_2 - y_1)^2} = \sqrt{(6-5)^2 + (3-5)^2} = \sqrt{(1)^2 + (-2)^2}$$

$$AS = \sqrt{5} \text{ units}$$

$$SE = \sqrt{(x_2 - x_1)^2 + (y_2 - y_1)^2} = \sqrt{(5-3)^2 + (5-4)^2} = \sqrt{(2)^2 + (1)^2}$$

$$SE = \sqrt{5} \text{ units}$$

The consecutive sides are congruent because the sides have the same distance.

4. **Yes**

$$\text{Midpoint of } \overline{SB} = \left( \frac{5+4}{2}, \frac{5+2}{2} \right) = (4.5, 3.5)$$

$$\text{Midpoint of } \overline{AE} = \left( \frac{6+3}{2}, \frac{3+4}{2} \right) = (4.5, 3.5)$$

The diagonals bisect each other because their midpoints are the same.

## EXERCISE 9-3

1. ***RHOM* is a rhombus.** Using the distance formula to find the lengths of all of the sides helps to classify the quadrilateral.

$$d = \sqrt{(x_2 - x_1)^2 + (y_2 - y_1)^2}$$

$$RH = \sqrt{(7-3)^2 + (1-(-1))^2}$$

$$RH = \sqrt{(4)^2 + (2)^2}$$

$$RH = \sqrt{20}$$

$$d = \sqrt{(x_2 - x_1)^2 + (y_2 - y_1)^2}$$
$$HO = \sqrt{(9-7)^2 + (5-1)^2}$$
$$HO = \sqrt{(2)^2 + (4)^2}$$
$$HO = \sqrt{20}$$

$$d = \sqrt{(x_2 - x_1)^2 + (y_2 - y_1)^2}$$
$$OM = \sqrt{(5-9)^2 + (3-5)^2}$$
$$OM = \sqrt{(-4)^2 + (-2)^2}$$
$$OM = \sqrt{20}$$

$$d = \sqrt{(x_2 - x_1)^2 + (y_2 - y_1)^2}$$
$$RM = \sqrt{(5-3)^2 + (3-(-1))^2}$$
$$RM = \sqrt{(2)^2 + (4)^2}$$
$$RM = \sqrt{20}$$

$RHOM$ is a rhombus because the sides are all equal in distance.

2. **(2, 15)**

$$\text{Midpoint} = \left( \frac{x_2 + x_1}{2}, \frac{y_2 + y_1}{2} \right)$$

$$\text{Midpoint} = \left( \frac{1+3}{2}, \frac{18+12}{2} \right) = (2,15)$$

3. **(10.5, −5)**

$$\text{Midpoint} = \left( \frac{x_2 + x_1}{2}, \frac{y_2 + y_1}{2} \right)$$

$$\text{Midpoint} = \left( \frac{17+4}{2}, \frac{-2+(-8)}{2} \right) = (10.5,-5)$$

**4. T(−8,−5)**   We were given $M(−2,10)$ is the midpoint of $\overline{AT}$, and point $A$ has coordinates (4,25). To find the $x$ and $y$ coordinates for point $T$, we will use the fact that the midpoint is the average of the $x$ and $y$ coordinates. This means the average of the $x$ coordinate for point $T$ and the $x$ coordinate for point $A$ equals the $x$ coordinate for the midpoint. We can set up the same equation to find the $y$ coordinate for point $T$. You can cross multiply to solve for the $x$ and $y$ coordinates; therefore

$$\frac{4+x}{2} = \frac{-2}{1}$$

$$\cancel{4}+x=-4$$

$$\underline{-\cancel{4} \qquad -4}$$

$$x=-8$$

$$\frac{25+y}{2} = \frac{10}{1}$$

$$\cancel{25}+y=20$$

$$\underline{-\cancel{25} \quad -25}$$

$$y=-5$$

**5.** $-\dfrac{1}{2}$

$$m\overline{WY} = \frac{y_2-y_1}{x_2-x_1} = \frac{18-15}{-3-3}$$

$$m\overline{WY} = \frac{3}{-6} = -\frac{1}{2}$$

**6. 3**

$$m\overline{CD} = \frac{y_2-y_1}{x_2-x_1} = \frac{12-(-6)}{7-1}$$

$$m\overline{CD} = \frac{18}{6} = 3$$

**7. −2**   Parallel lines have equal slopes. Once the equation is in $y = mx + b$ form, the coefficient for the $x$ represents the slope of the line.

**8.**  $-\dfrac{2}{3}$

Solve for $y$ first: $y = \dfrac{3}{2}x + 2.5$. The slope of this line is $\dfrac{3}{2}$.

Perpendicular lines have slopes that are negative reciprocals;

therefore, the slope of the line perpendicular to the given line is $-\dfrac{2}{3}$.

**9.** $y = \dfrac{1}{4}x + 16$

Solve for $y$: $y = -4x + 5$.

The slope of the given line is $-4$. Perpendicular lines have slopes that are

negative reciprocals therefore the slope of the new line must be $\dfrac{1}{4}$.

We can substitute the $\dfrac{1}{4}$ and the point (16, 20) into the point-slope form

of a line to get the equation of the new line.

Let $(x_1, y_1) = (16, 20)$

Let $m = \dfrac{1}{4}$

$y - y_1 = m(x - x_1)$

$y - 20 = \dfrac{1}{4}(x - 16)$

$y - \cancel{20} = \dfrac{1}{4}x - 4$

$\underline{\quad + \cancel{20} \qquad + 20 \quad}$

$\qquad y = \dfrac{1}{4}x + 16$

10. **$y=-2x+30$**

Solve for y: $y=-2x+7$.

The slope of the given line is $-2$. Parallel lines have equal slopes; therefore, the slope of the new line must also be $-2$.

We can substitute the $-2$ and the point $(10, 10)$ into the point-slope form of a line to get the equation of the new line.

Let $(x_1,y_1)=(10,10)$

Let $m=-2$

$$y-y_1=m(x-x_1)$$
$$y-10=-2(x-10)$$
$$y-10=-2x+20$$
$$\underline{+10\qquad\quad +10}$$
$$y=-2x+30$$

11. **(4,23)**   The ratio 3:1 means that the line segment is being divided in such a way that $\overline{PA}$ contains three congruent parts and $\overline{AS}$ contains one congruent part for a total of four congruent parts. We are looking to find the location of point $A$ such that $\overline{PA}$ is 3/4 of the total distance of $\overline{PS}$.

**Finding the $x$-value of point $A$:**

We want the horizontal distance from point $P$ to point $A$ to be 3/4 of the horizontal distance from point $P$ to point $S$. The horizontal distance is the difference in the $x$ values of the endpoints:

$$x_2-x_1$$
$$\downarrow$$
$$6--2=8$$

The horizontal distance from point $P$ to point $A$ must be 3/4 of 8: $\dfrac{3}{4}(8)=6$

We need the $x$ value for point $A$ to be 6 units to the right of the $x$ value for point $P$:

$$-2 + 6 = 4$$

The $x$ value for point $A$ is 4.

**Finding the y-value of point A:**

We want the vertical distance from point $P$ to point $A$ to be 3/4 of the vertical distance from point $P$ to point $S$. The vertical distance is the difference in the $y$ values of the two endpoints:

$$y_2 - y_1$$
$$\downarrow$$
$$29 - 5 = 24$$

The vertical distance from point $P$ to point $A$ must be $\dfrac{3}{4}$ of 24: $\dfrac{3}{4}(24) = 18$.

We need the $y$ value for point $A$ to be 18 units above the $y$ value for point $P$:

$$5 + 18 = 23$$

The $y$ value for point $A$ is 23.

12. **(7,15)**   The ratio 2:1 means that the line segment is being divided in such a way that $\overline{CT}$ contains two congruent parts and $\overline{TD}$ contains one congruent part for a total of three congruent parts. We are looking to find the location of point $T$ such that $\overline{CT}$ is 2/3 of the total distance of $\overline{CD}$.

**Finding the x-value of point T:**

We want the horizontal distance from point $T$ to point $C$ to be 2/3 of the horizontal distance from points $C$ to $D$. The horizontal distance is the difference in the $x$ values of the endpoints:

$$x_2 - x_1$$
$$\downarrow$$
$$8 - 5 = 3$$

The horizontal distance from point $C$ to point $T$ must be 2/3 of 3: $\dfrac{2}{3}(3)=2$

We need the $x$ value for point $T$ to be two units to the right of the $x$ value for point $C$:

$5+2=7$

The $x$ value for point $T$ is 7.

**Finding the $y$ value of point $T$:**

We want the vertical distance from point $C$ to point $T$ to be 2/3 of the vertical distance from points $C$ to point $D$. The vertical distance is the difference in the $y$ values of the two endpoints:

$$y_2 - y_1$$
$$\downarrow$$
$$17 - 11 = 6$$

The vertical distance from point $C$ to point $T$ must be 2/3 of 6: $\dfrac{2}{3}(6)=4$

We need the $y$ value for point $T$ to be four units above the $y$ value for point $C$

$11+4=15$

The $y$ value for point $T$ is 15.

# Transformations

## EXERCISE 10-1

1. **(4,−25)**   When a point is reflected over the $x$-axis, the rule is to negate the $y$ value.

2. **(−4,25)**   When a point is reflected over the $y$-axis, the rule is to negate the $x$-value.

3. **(25,4)**   When a point is reflected over the line $y = x$, the rule is to switch the $x$ and $y$ coordinates.

4. **(6,25)**   $x = 5$ is a vertical line. $P(4,25)$ is one unit to the left of it. Therefore, the mirror image is one unit to the right of $x = 5$.

5. **(4,15)**  $y = 20$ is a horizontal line. $P(4,25)$ is five units above it. Therefore the mirror image is five units below $y = 20$.

6. **(12,24)**  $T_{8-1}$ represents a translation with the rule: $(x + 8, y - 1)$.

7. **(12,75)**  $D_3$ represents a dilation with a scale factor of 3. Multiply the $x$ and $y$ coordinates by 3.

8. **(1,6.25)**  $D_{\frac{1}{4}}$ represents a dilation with a scale factor of $\frac{1}{4}$. Multiply the $x$ and $y$ coordinates by $\frac{1}{4}$.

## EXERCISE 10-2

1. **(−12,2)**  When a point is rotated $90°$ about the origin, the rule is to negate the $y$ coordinate and switch it with the $x$ coordinate.

2. **(−3,−2)**  A *clockwise* rotation of $90°$ is equivalent to a counterclockwise rotation of $270°$. When a point is rotated $270°$ around the origin, the rule is to negate the $x$ coordinate and switch it with the $y$ coordinate.

## EXERCISE 10-3

1. **False**  A dilation does *not* preserve distance, which means it is *not* a rigid motion.

2. **True**  A translation preserves distance (size) and orientation (order).

3. **False**  A line reflection *is* a rigid motion because it preserves distance but it *does not* preserve orientation because the order changes.

## EXERCISE 10-4

1. **$x = 1$**  $x = 1$ is the line of symmetry for the two triangles.

2. **$C'(9,1)$ $D'(16,−6)$**  $T_{4,-9}$ represents a translation with the rule: $(x + 4, y - 9)$.

3. **(−3, 8)**  $D_{\frac{1}{5}}$ represents a dilation with a scale factor of $\frac{1}{5}$. Multiply the $x$ and $y$ coordinates by $\frac{1}{5}$.

4. **1.5**  Divide the image $x$ and $y$ coordinates by the pre-image $x$ and $y$ coordinates to find the scale factor. $\frac{6}{4} = 1.5$ and $\frac{9}{6} = 1.5$

5. **(−15,35)** Divide the image $x$ and $y$ coordinates by the pre-image $x$ and $y$ coordinates to find the scale factor. $\dfrac{10}{-2} = -5$ and $\dfrac{-50}{10} = -5$. Multiply the $x$ and $y$ coordinates for point $B$ by $-5$ to find its image point.

6. **$H'(11,9)\ I'(13,1)$** The composition is performed from right to left.

$$H(3,5) \xrightarrow[D_2]{} (6,10) \xrightarrow[T_{5,-1}]{} H'(11,9)$$

$$I(4,1) \xrightarrow[D_2]{} (8,2) \xrightarrow[T_{5,-1}]{} I'(13,1)$$

7. **(5,5)** The composition is performed from right to left.

$$P(-4,3)\xrightarrow[r_{y=x}]{}(3,-4)\xrightarrow[R_{90°}]{}(4,3)\xrightarrow[T_{1,2}]{}P'(5,5)$$

8. **(3,−3) invariant** This is an invariant point because it lies on the line of reflection. The pre-image and image are in the exact same location.

9. **$r_{y-axis} \circ D_3(\triangle AMZ)$** The sequence of transformations is a dilation with a scale factor of 3, followed by a reflection over the $y$-axis.

10. **(2,0)** Place your pencil at point $K(4,2)$. To get to point $G(3,1)$ from point $K$, go left one unit and down one unit. Repeat this pattern again for a total of TWO times since the scale factor is 2. So, from $G(3,1)$, go left one and down one. Arrive at $G'(2,0)$.

# 11

# Circle Theorems Involving Angles and Segments

## EXERCISE 11-1

1. **18** A radius and a tangent are perpendicular at their point of tangency making triangle $ABO$ a right triangle. We were given $AO = 15$ and $AB = 12$ and can find the radius of the circle using the Pythagorean theorem. The diameter of a circle is twice the length of the radius; therefore

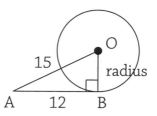

$$(12)^2 + (OB)^2 = (15)^2$$
$$\cancel{144} + (OB)^2 = 225$$
$$\underline{-\cancel{144} \qquad\quad -144}$$
$$\sqrt{(OB)^2} = \sqrt{81}$$
$$OB = 9$$
$$\text{diameter} = 2(9) = 18$$

2. **70**   We were given $SG = 12$, $AI = 8$, and $TN = 15$, which means $GI = 12$, $AN = 8$, and $TS = 15$ because tangents drawn from the same external point are congruent. The perimeter of a triangle is equal to the sum of all the sides of the triangle; therefore

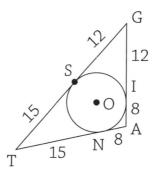

$$15 + 15 + 8 + 8 + 12 + 12 = 70$$

## EXERCISE 11-2

1. **12**   To solve for $SN$ we need to use the formula

$$(tangent)^2 = (exterior\ part\ of\ secant)\ (entire\ secant)$$

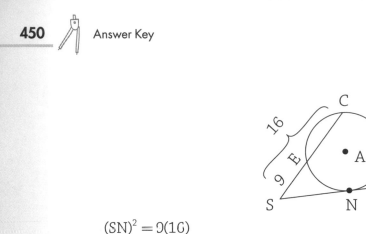

$$(SN)^2 = 9(16)$$
$$\sqrt{(SN)^2} = \sqrt{144}$$
$$SN = 12$$

**2. 3** To solve for $SE$, we need to use the formula

$(tangent)^2 = (exterior\ part\ of\ secant)\ (entire\ secant)$

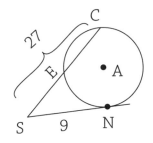

$$(SN)^2 = (SE)(SC)$$
$$(9)^2 = (SE)(27)$$
$$\frac{81}{27} = \frac{\cancel{27}(SE)}{\cancel{27}}$$
$$3 = SE$$

**3. 6** To solve for $CE$, we need to use the formula

$(tangent)^2 = (exterior\ part\ of\ secant)\ (entire\ secant)$

Let $x = CE$.

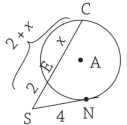

$$(SN)^2 = (SE)(SC)$$
$$(4)^2 = 2(2+x)$$
$$16 = \cancel{4} + 2x$$
$$\frac{-4 \quad -\cancel{4}}{\dfrac{12}{2} = \dfrac{\cancel{2}x}{\cancel{2}}}$$
$$6 = x$$
$$6 = CE$$

4. **16**    To solve for $SN$, we need to use the formula

$$(tangent)^2 = (exterior\ part\ of\ secant)\ (entire\ secant)$$

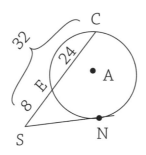

$$(SN)^2 = (SE)(SC)$$
$$(SN)^2 = 8(32)$$
$$\sqrt{(SN)^2} = \sqrt{256}$$
$$SN = 16$$

## EXERCISE 11-3

1. **75°**    The measure of a central angle is equal to the measure of the intercepted arc.

2. **25°**    The measure of a central angle is equal to the measure of the intercepted arc, which is why arc $m\overset{\frown}{AC} = 50°$. The measure of an inscribed angle is equal to half the measure of the intercepted arc, which is why $m\angle CBA = 25°$. This is shown in the accompanying diagram.

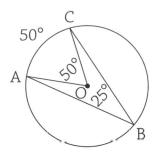

3. **140°** The measure of an inscribed angle is equal to half the measure of the intercepted arc. We were given $m\angle ABC = 40°$, which means the intercepted arc measures 80°. If two chords of a circle are congruent, then their intercepted arcs are congruent. The sum of the arcs of a circle is 360° therefore,

Let $x = m\overset{\frown}{AB}$.

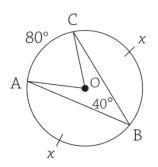

$$x + x + 80 = 360°$$
$$2x + 80 = 360$$
$$\underline{-80 \quad -80}$$
$$\frac{2x}{2} = \frac{280}{2}$$
$$x = 140$$
$$m\overset{\frown}{AB} = 140°$$

### EXERCISE 11-4

1. **8** When two chords intersect in a circle, the product of the segments of one chord is equal to the product of the segments of the other chord. We were given $LY = 4$, $CY = 10$, and $KY = 5$; therefore

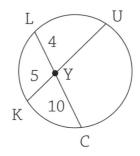

$$(KY)(UY) = (LY)(CY)$$
$$5(UY) = 4(10)$$
$$\frac{\cancel{5}(UY)}{\cancel{5}} = \frac{40}{5}$$
$$UY = 8$$

2. **8**   When two chords intersect in a circle, the product of the segments of one chord is equal to the product of the segments of the other chord. We were given $UY = 6$, $KY = 4$, and $CY$ is 5 more than $LY$; therefore

Let $x = LY$.

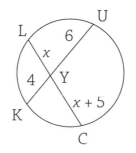

$$(KY)(UY) = (LY)(CY)$$
$$4(6) = x(x+5)$$
$$\cancel{24} = x^2 + 5x$$
$$\underline{-\cancel{24} \quad -24}$$
$$0 = x^2 + 5x - 24$$
$$0 = \frac{(x+8)(x-3)}{\cancel{x = -8} \; x = 3}$$
$$CY = x + 5$$
$$CY = 3 + 5$$
$$CY = 8$$

3. **80°** When two chords intersect in a circle, they create vertical angles that are equal to the average of the arcs they intercept. We were given $m\overset{\frown}{LU} = 72°$ and $m\overset{\frown}{KC} = 88°$; therefore

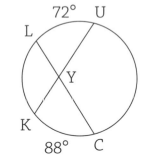

$$m\angle LYU = \frac{m\overset{\frown}{LU} + m\overset{\frown}{KC}}{2}$$

$$m\angle LYU = \frac{72 + 88}{2}$$

$$m\angle LYU = \frac{160}{2}$$

$$m\angle LYU = 80°$$

4. **117°** When two chords intersect in a circle, they create vertical angles that are equal to the average of the arcs they intercept. We were given $m\overset{\frown}{KL} = 47°$ and $m\angle LYK = 82°$; therefore

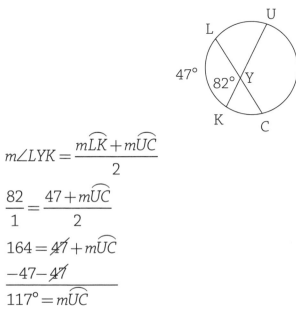

$$m\angle LYK = \frac{m\overset{\frown}{LK} + m\overset{\frown}{UC}}{2}$$

$$\frac{82}{1} = \frac{47 + m\overset{\frown}{UC}}{2}$$

$$164 = \cancel{47} + m\overset{\frown}{UC}$$

$$\frac{-47 - \cancel{47}}{117° = m\overset{\frown}{UC}}$$

5. **95°**   We can solve for the missing arc because the sum of all the arcs of a circle is 360°. This is important to us because finding the measure of a vertical angle requires knowing the measure of the two arcs the vertical angles open up to. When two chords intersect in a circle, they create vertical angles that are equal to the average of the arcs they intercept. We were given $m\overset{\frown}{CK}=100°$, $m\overset{\frown}{KL}=60°$, and $m\overset{\frown}{LU}=70°$; therefore

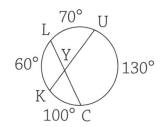

$$70+60+100+m\overset{\frown}{CU}=360$$

$$230+m\overset{\frown}{CU}=360$$

$$\underline{-230 \qquad -230}$$

$$m\overset{\frown}{CU}=130°$$

$$m\angle CYU=\frac{m\overset{\frown}{KL}+m\overset{\frown}{CU}}{2}$$

$$m\angle CYU=\frac{60+130}{2}$$

$$m\angle CYU=\frac{190}{2}$$

$$m\angle CYU=95°$$

## EXERCISE 11-5

1. **37°**   The measure of an exterior angle is equal to half the difference of the measures of the intercepted arcs. We were given $m\overset{\frown}{AO}=56°$ and $m\overset{\frown}{CD}=130°$; therefore

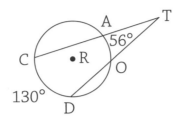

$$m\angle T = \frac{1}{2}(m\widehat{CD} \quad m\widehat{AO})$$

$$m\angle T = \frac{1}{2}(130° - 56°)$$

$$m\angle T = \frac{1}{2}(74)$$

$$m\angle T = 37°$$

2. **158°**  The measure of an exterior angle is equal to half the difference of the measures of the intercepted arcs. We were given $m\widehat{AO} = 68°$ and $m\angle T = 45°$; therefore

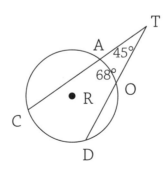

$$m\angle T = \frac{1}{2}(m\widehat{CD} - m\widehat{AO})$$

$$\left(\frac{2}{1}\right)45 = \frac{1}{2}(m\widehat{CD} - 68)\left(\frac{2}{1}\right)$$

$$90 = m\widehat{CD} - 68$$

$$\underline{+68 \qquad +68}$$

$$158° = m\widehat{CD}$$

**3. 6**  To solve for *TA*, we need to use the formula (Exterior part of secant 1) (Entire secant 1) = (Exterior part of secant 2)(Entire secant 2). We were given $DT = 12$, $OT = 8$, and $CT = 16$; therefore

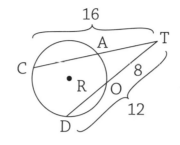

$$(TA)(TC) = (TO)(TD)$$
$$(TA)(16) = 8(12)$$
$$\frac{\cancel{16}(TA)}{\cancel{16}} = \frac{96}{16}$$
$$TA = 6$$

# Circumference and the Area of Circles

### EXERCISE 12-1

1. **615.75 sq units**

   $A = \pi r^2$

   $A = \pi(14)^2 = 615.75$ sq units

2. **15.7 units**

   $C = \pi d$

   $C = \pi(5)$

   $C = 15.7$ units

3. **18 ft**

$$A = \pi r^2$$

$$\frac{81\pi}{\pi} = \frac{\pi r^2}{\pi}$$

$$\sqrt{81} = \sqrt{r^2}$$

$$9 \text{ft} = r$$

The radius is 9 ft; therefore, the diameter is 18 ft.

### EXERCISE 12-2

1. $\dfrac{640\pi}{9} u^2$

$$A = \frac{\theta}{360°}(\pi r^2)$$

$$A = \frac{100°}{360°}(\pi (16)^2)$$

$$A = \frac{100°}{360°}(256\pi)$$

$$A = \frac{640\pi}{9} u^2$$

2. $\dfrac{80\pi}{9}$ **units**

$$\text{Arc length} = \frac{\theta}{360°}(2\pi r)$$

$$\text{Arc length} = \frac{100°}{360°}(2\pi(16))$$

$$\text{Arc length} = \frac{100°}{360°}(32\pi)$$

$$\text{Arc length} = \frac{80\pi}{9} \text{ units}$$

3. $\dfrac{80\pi}{9}+32\,\textbf{units}$

    Perimeter of the Sector $=\dfrac{100°}{360°}(2\pi(16))+2(16)$

    Perimeter of the Sector$=\dfrac{80\pi}{9}+32$ units

## EXERCISE 12-3

1. **Standard Form**    The standard form of a circle is $(x-h)^2+(y-k)^2=r^2$.

2. **(5, −10)**    The standard form of a circle is $(x-h)^2+(y-k)^2=r^2$. The location $(h,k)$ represents the center of the circle.

3. $2\sqrt{5}$    The standard form of a circle is $(x-h)^2+(y-k)^2=r^2$.

    $r^2=20$

    $r=\sqrt{20}=2\sqrt{5}$

4.     $x^2+y^2-10x+20y+106=0$

    $(x-5)(x-5)+(y+10)(y+10)=20$

    $x^2-10x+25+y^2+20y+100=20$

    $x^2+y^2-10x+20y+125=\cancel{20}$

$$\underline{\phantom{x^2+y^2-10x+20y+125=20}\; -20-\cancel{20}}$$

    $x^2+y^2-10x+20y+105=0$

## EXERCISE 12-4

1. $x^2+y^2=1$    The standard form of a circle is $(x-h)^2+(y-k)^2=r^2$.

    $(x-0)^2+(y-0)^2=(1)^2$

    $x^2+y^2=1$

2. $x^2+\left(y-\dfrac{4}{3}\right)^2=17$    The standard form of a circle is $(x-h)^2+(y-k)^2=r^2$.

$$(x-0)^2 + \left(y - \frac{4}{3}\right)^2 = (\sqrt{17})^2$$

$$x^2 + \left(y - \frac{4}{3}\right)^2 = 17$$

3. $(x+2)^2 + (y+5)^2 = 9$   The standard form of a circle is $(x-h)^2 + (y-k)^2 = r^2$ where $(h,k)$ represents the center of the circle and $r$ represents the radius of the circle. The radius of a circle is half the diameter.

$$(x+2)^2 + (y+5)^2 = (3)^2$$

$$(x+2)^2 + (y+5)^2 = 9$$

4. We were given a circle whose equation is $(x+3)^2 + (y+1)^2 = 16$. The standard form of a circle is $(x-h)^2 + (y-k)^2 = r^2$ where $(h,k)$ represents the center of the circle and $r$ represents the radius of the circle. The center of this circle is therefore $(-3,-1)$ and the radius is 4. Plot the center on the coordinate grid. Place four points on the graph for the circle to pass through. Each point should be four units away from the center of the circle (above, below, to the left, and to the right), as shown in the following diagram.

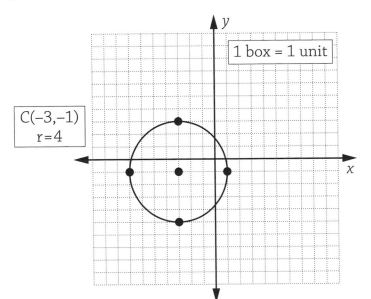

## EXERCISE 12-5

1. **The point lands on the circle.**

$$(-2+4)^2 + (5-1)^2 = 20$$
$$20 \qquad = 20$$

2. **Exterior point.**

$$(0+4)^2 + (8-1)^2 > 20$$
$$65 \qquad > 20$$

3. **Interior point.**

$$(-5+4)^2 + (3-1)^2 < 20$$
$$5 \qquad < 20$$

## EXERCISE 12-6

1. **Center: (2,−6); radius: 7**   To find the center and the radius, you must complete the square to get the equation written in standard form.

$$(x^2 - 4x) + (y^2 + 12y) = 9$$
$$(x^2 - 4x + 4) + (y^2 + 12y + 36) = 9 + 4 + 36$$
$$(x-2)(x-2) + (y+6)(y+6) = 49$$
$$(x-2)^2 + (y+6)^2 = 49$$
$$\text{Center} : (2,-6)$$
$$\sqrt{r^2} = \sqrt{49}$$
$$r = 7$$

2. **160°**

$$A = \frac{\theta}{360}(\pi r^2)$$

$$144\pi = \frac{\theta}{360}(\pi(18)^2)$$

$$\left(\frac{360}{1}\right)144\pi = \frac{\theta}{\cancel{360}}(324\pi)\left(\frac{\cancel{360}}{1}\right)$$

$$\frac{51,840\pi}{324\pi} = \frac{\cancel{324\pi}}{\cancel{324\pi}}\theta$$

$$160° = \theta$$

3. **56.6 meters**

$$A = \frac{\theta}{360}(\pi r^2)$$

$$\left(\frac{360}{45}\right)400\pi = \frac{\cancel{45}}{\cancel{360}}(\pi r^2)\left(\frac{\cancel{360}}{\cancel{45}}\right)$$

$$\frac{3,200\cancel{\pi}}{\cancel{\pi}} = \frac{\cancel{\pi}r^2}{\cancel{\pi}}$$

$$\sqrt{3,200} = \sqrt{r^2}$$

$$56.6 = r$$

# Volume of Three-Dimensional Shapes

**EXERCISE 13-1**

1. **1,760$\pi$ in²**   The formula for finding the surface area of a cylinder is $SA = 2\pi rh + 2\pi r^2$. We were given that the radius of the base of the cylinder is 20 inches and the height is 24 inches; therefore

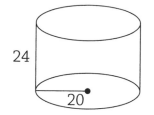

$$SA = 2\pi r^2 + 2\pi rh$$
$$SA = 2\pi(20)^2 + 2\pi(20)(24)$$
$$SA = 800\pi + 960\pi$$
$$SA = 1{,}760\pi \text{ in}^2$$

2. **2,199.1 ft³**   The formula for finding the volume of a cylinder is $V = \pi r^2 h$. We were given that the diameter of the cylinder is 20 feet, which means the radius is 10 feet and the height of the cylinder is 7 feet; therefore

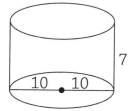

$$V = Bh$$
$$V = \pi r^2 h$$
$$V = \pi(10)^2(7)$$
$$V = 2{,}199.1 \text{ ft}^3$$

3. **4**   The formula for finding the volume of a cylinder is $V = \pi r^2 h$. We were given that the volume of the cylinder is $100\pi$ and the radius is 5; therefore

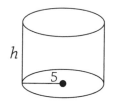

$$V = Bh$$
$$V = \pi r^2 h$$
$$100\pi = \pi(5)^2 h$$
$$\frac{100\pi}{25\pi} = \frac{25\pi h}{25\pi}$$
$$4 - h$$

4. **32.7 in³**   The formula for finding the volume of a right circular cone is $V = \frac{1}{3}\pi r^2 h$. We were given that the height of the cone is 5 inches and the radius is 2.5 inches; therefore

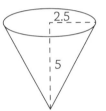

$$V = \frac{1}{3}Bh$$

$$V = \frac{1}{3}\pi r^2 h$$

$$V = \frac{1}{3}\pi(2.5)^2(5)$$

$$V = 32.7\, \text{in}^3$$

5. **5 inches**   The slant height of a right circular cone can be found using the Pythagorean theorem. We were given that the radius of the cone is 3 inches and the height is 4 inches; therefore

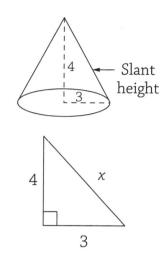

$$a^2 + b^2 = c^2$$
$$(3)^2 + (4)^2 = x^2$$
$$\sqrt{25} = \sqrt{x^2}$$
$$5 = x$$

6. **24 cm**   The slant height of a right circular cone is the hypotenuse of the right triangle containing the radius and height of the cone. The height can be found using the Pythagorean theorem. We were given that the slant height of the cone is 25 centimeters and the radius is 7 centimeters; therefore

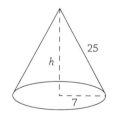

$$a^2 + b^2 = c^2$$
$$(h)^2 + (7)^2 = (25)^2$$
$$h^2 + 49 = 625$$
$$\underline{-49 \quad -49}$$
$$\sqrt{h^2} = \sqrt{576}$$
$$h = 24 \text{ cm}$$

7. **48 m³**   The volume of a pyramid can be found using the formula $V = \frac{1}{3}Bh$, where $B$ is the area of the base of the pyramid and $h$ is the height of the pyramid. We were given that the base of the square pyramid is 4 meters long and its height is 9 meters; therefore

$$V = \frac{1}{3}Bh$$

$$V = \frac{1}{3}l \cdot w \cdot h$$

$$V = \frac{1}{3}(4)(4)(9)$$

$$V = 48 \, \text{m}^3$$

8. **180 in³**   The volume of a rectangular prism can be found using the formula $V = l \cdot w \cdot h$. We were given a 6-inch-high prism whose base is a rectangle measuring 10 inches by 3 inches; therefore

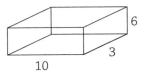

$$V = Bh$$

$$V = l \cdot w \cdot h$$

$$V = (10)(3)(6)$$

$$V = 180 \, \text{in}^3$$

9. **144 units³**   The volume of a triangular prism can be found using the formula $V = \frac{1}{2}$ (Base of triangle)(Height of triangle)(Height of prism).

We were given the triangular base is 6 units long, the height of the triangle is 4 units, and the rectangle is 12 units wide; therefore

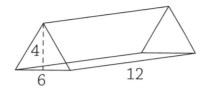

$$V = Bh$$

$$V = \frac{1}{2}(\Delta \text{base})(\Delta \text{height})(h)$$

$$V = \frac{1}{2}(6)(4)(12)$$

$$V = 144 \text{ units}^3$$

10. **179.59 cm³**  The volume of a sphere can be found using the formula $V = \frac{4}{3}\pi r^3$. We were given that the diameter is 7 cm, which means the radius is 3.5 cm; therefore

$$V = \frac{4}{3}\pi r^3$$

$$V = \frac{4}{3}\pi (3.5)^3$$

$$V = 179.59 \text{ cm}^3$$

11. **1.42 cm**   The volume of a sphere can be found using the formula
$V = \dfrac{4}{3}\pi r^3$. We were given that the sphere has a volume of 12 cm³; therefore

$$V = \frac{4}{3}\pi r^3$$

$$\left(\frac{3}{4}\right)12 = \frac{\cancel{4}}{\cancel{3}}\pi r^3 \left(\frac{\cancel{3}}{\cancel{4}}\right)$$

$$\frac{9}{\pi} = \frac{\cancel{\pi} r^3}{\cancel{\pi}}$$

$$\sqrt[3]{\frac{9}{\pi}} = \sqrt[3]{r^3}$$

$$1.42 = r$$

## EXERCISE 13-2

1. **735π units³**   When a rectangle is rotated around one of its sides, it
results in a cylinder with $\overline{AB}$ as the radius and $\overline{AD}$ as the height. The
formula for finding the volume of a cylinder is $V = \pi r^2 h$; therefore

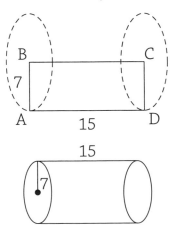

$V = Bh$

$V = \pi r^2 h$

$v = \pi (7)^2 (15)$

$V = 735\pi$ cubic units

2. **3,769.9 cm³**  When a rectangle is rotated around its perpendicular bisector, it results in a cylinder with half of $\overline{BC}$ as the radius and $\overline{BA}$ as the height. The formula for finding the volume of a cylinder is $V = \pi r^2 h$; therefore

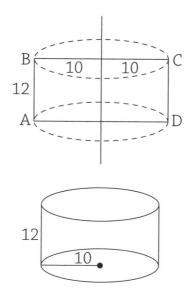

$V = Bh$

$V = \pi r^2 h$

$V = \pi (10)^2 (12)$

$V = 3,769.9 \text{ cm}^3$

3. **209.4 units³**  When a right triangle is rotated around one of its sides, it results in a right circular cone with $\overline{AC}$ as the radius and $\overline{BC}$ as the height. The formula for the volume of a right circular cone is $V = \dfrac{1}{3}\pi r^2 h$; therefore

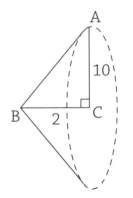

$$V = \frac{1}{3}Bh$$

$$V = \frac{1}{3}\pi r^2 h$$

$$V = \frac{1}{3}\pi(10)^2(2)$$

$$V = 209.4 \text{ units}^3$$

**4. 904.78 in³** When a semicircle is rotated around its diameter, it results in a sphere whose radius is the same as the radius of the semicircle. The formula for the volume of a sphere is $V = \frac{4}{3}\pi r^3$; therefore

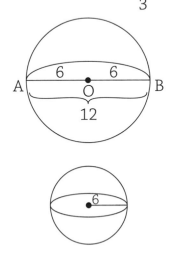

$$V = \frac{4}{3}\pi r^3$$

$$V = \frac{4}{3}\pi(6)^3$$

$$V = 904.78 \text{ in}^3$$

# 14
## Constructions

### EXERCISE 14-1

1. The following diagram shows the arcs necessary to construct the perpendicular bisector of $\overline{AB}$.

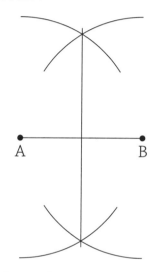

2. The following diagram shows the arcs necessary to construct an angle congruent to a given angle.

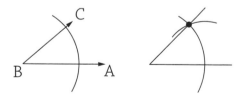

3. The following diagram shows the arcs necessary to construct a segment congruent to a given segment.

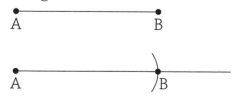

4. The following diagram shows the arcs necessary to construct an equilateral triangle with side length $\overline{AB}$.

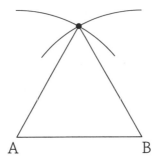

5. The following diagram shows the arcs necessary to construct the angle bisector for $\angle ABC$.

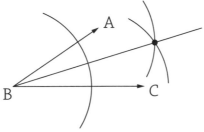

6. The following diagram shows the arcs necessary to construct a line perpendicular to $\overline{AB}$ through point $C$.

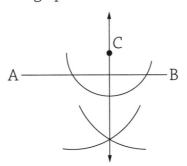

7. The following diagram shows the arcs necessary to construct a hexagon inscribed in a circle.

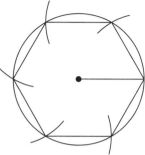

8. The following diagram shows the arcs necessary to construct $\overline{BD}$, the median of $\triangle ABC$.

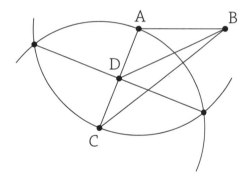

9. The following diagram shows the arcs necessary to construct the rotation of point $P$ 60° clockwise around point $C$. This construction first required the construction of a 60° angle.

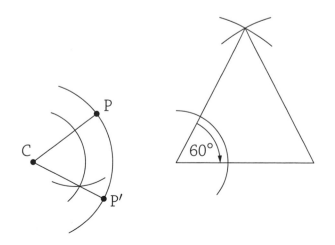

10. **Angle bisector** The construction of an angle bisector can be easily recognized because it will always have an arc drawn through the angle and then two smaller arcs intersecting with a ray drawn through the angle.

11. **Perpendicular bisector** The construction of a perpendicular bisector can be easily recognized because it will always have a pair of arcs intersecting above the segment and a pair of arcs intersecting below the segment.

12. The following diagram shows the arcs necessary to construct the line of reflection for $\triangle ABC$, and its image $\triangle A'B'C'$.

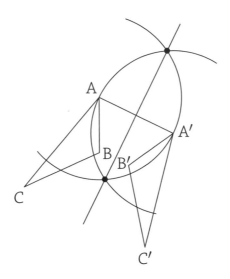

13. Construct perpendicular lines to create right angles. Bisect a right angle to create a 45°angle.

14. The following diagram shows the arcs necessary to construct a triangle congruent to the given triangle.

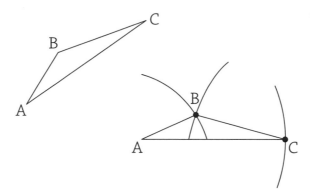

15. The following diagram shows the arcs necessary to construct the dilation of $\overline{AB}$ by a scale factor of 2 with the center of dilation at point $C$.

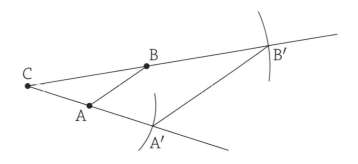

# Teacher's Guide

**T**eaching geometry can be a bit overwhelming at times. There are many obstacles that present challenges to students who strive to master the course. Three specific areas that prove to be challenging for students include: working with three-dimensional figures, the dexterity of using a compass, and completing geometric proofs. We have provided strategies to overcome these challenges and help make your geometry classroom a place for student success!

## 3D Figures

Working with 3D figures can be challenging since many students struggle to think and visualize in a three-dimensional manner. Use concrete examples and bring visual aids into the classroom. A soccer ball, a shoebox, a soup can, a toilet paper roll, and an ice cream cone can all be useful tools to help students obtain conceptual and true understanding of the 3D shapes. Bring enough objects so that students can have the opportunity to touch, feel, and study an item in order to help them solve volume and surface area problems. As they become more comfortable with concrete examples, they can move towards thinking more abstractly. With time and practice, they will not need to have the object in their hand and, instead, will be able to visualize concepts in their head.

## Using a Compass

Most students have never even held a compass in their hand before they entered your classroom, let alone used it to mark construction arcs.

Therefore, you shouldn't introduce the compass with construction questions. It takes time and practice for students to understand how to hold and use a compass correctly, and for many students it can get very frustrating. Instead, have students draw circle art and encourage them to be creative with the compass to build proficiency. Once they have the feel for the compass and can make circles successfully without the compass slipping or changing size, it is time to start introducing constructions.

## Geometric Proofs

Success with proofs requires students to be fluent with the academic vocabulary first. Once students are proficient with the vocabulary, the proofs will be easier to tackle. Therefore, you should pre-teach the vocabulary and spend time ensuring that students have a good understanding of the words. Have students use flash cards or even create games to practice the information that vocabulary words, such as perpendicular bisector, can give them before jumping into doing a full proof. Have students simply practice marking information in a diagram before setting up a full proof. Once your students have demonstrated that they have mastered the vocabulary, it is time to introduce it in a proof. Start small! Remember, they are training their brain to solve a visual puzzle. It takes time!

## Helpful Resources

**Khan academy** (Khanacademy.org) has wonderful videos students can watch at home to reteach, review, and practice what they have learned in the classroom.

**Delta math** (Deltamath.com) has a wide array of practice questions that you can assign to the students for homework, extra practice, or even quizzes. It shows solutions for similar examples if students are stuck on a problem, and it also gives students the option to watch a help video if they are stuck or just need a little more review on the topic.

# NOTES

**NOTES**

# NOTES

**NOTES**

**NOTES**

# NOTES

# NOTES

**NOTES**